Barcode in Back

The Globalization of Advertising

The role of advertising in everyday life and as a major employer in post-industrial economies is intimately bound up with processes of contemporary globalization. At the centre of the advertising industry are the global advertising agencies, which have an important role in developing global brands both nationally and internationally. This book identifies and addresses questions on the globalization of advertising through detailed study of the contemporary advertising industry in Detroit, Los Angeles and New York City and the way advertising work has changed in the three cities over recent years.

The Globalization of Advertising draws upon previously unpublished research to unpack the contemporary structure, spatial organization and city geographies of global advertising agencies. The book demonstrates how teamwork in contemporary advertising agencies, intra-organizational power relations and the distribution of organizational capabilities all define how global agencies operate as transnationally integrated organizations. This in turn allows understanding to be developed of the role of the offices of global agencies located in the three case study cities. The role of these three cities as preeminent markets for advertising in the USA is shown to have changed radically over recent years, experiencing both growth and decline in employment as a result of their position in global networks of advertising work; networks that operate in the context of a changing US economy and the rise of new and emerging centres of advertising in Asia and South America.

This book offers a cutting edge overview of recent and current trends in the globalization of advertising and new insights into the way global advertising agencies operate in and through world cities. It will be a valuable resource for researchers and students studying Geography, Management and Sociology.

James R. Faulconbridge is a Senior Lecturer in Economic Geography at Lancaster University, UK. **Jonathan V. Beaverstock** is Professor of Economic Geography at the University of Nottingham, UK. **Corinne Nativel** is a Lecturer in English and Economic Geography at the University of Franche-Comté in Besançon, France and a member of the CREW (Centre for Research on the English Speaking World) at the University of Paris III-Sorbonne Nouvelle. **Peter J. Taylor** FBA is Director of Globalization and World Cities Research Network (GaWC) and Professor of Geography at Northumbria University, UK.

Routledge studies in human geography

This series provides a forum for innovative, vibrant and critical debate within Human Geography. Titles will reflect the wealth of research that is taking place in this diverse and ever-expanding field. Contributions will be drawn from the main sub-disciplines and from innovative areas of work that have no particular sub-disciplinary allegiances.

Published:

1 **A Geography of Islands**
Small island insularity
Stephen A. Royle

2 **Citizenships, Contingency and the Countryside**
Rights, culture, land and the environment
Gavin Parker

3 **The Differentiated Countryside**
Jonathan Murdoch, Philip Lowe, Neil Ward and Terry Marsden

4 **The Human Geography of East Central Europe**
David Turnock

5 **Imagined Regional Communities**
Integration and sovereignty in the global south
James D. Sidaway

6 **Mapping Modernities**
Geographies of Central and Eastern Europe 1920–2000
Alan Dingsdale

7 **Rural Poverty**
Marginalisation and exclusion in Britain and the United States
Paul Milbourne

8 **Poverty and the Third Way**
Colin C. Williams and Jan Windebank

9 **Ageing and Place**
Edited by Gavin J. Andrews and David R. Phillips

10 **Geographies of Commodity Chains**
Edited by Alex Hughes and Suzanne Reimer

11 Queering Tourism
Paradoxical performances at Gay Pride parades
Lynda T. Johnston

12 Cross-Continental Food Chains
Edited by Niels Fold and Bill Pritchard

13 Private Cities
Edited by Georg Glasze, Chris Webster and Klaus Frantz

14 Global Geographies of Post Socialist Transition
Tassilo Herrschel

15 Urban Development in Post-Reform China
Fulong Wu, Jiang Xu and Anthony Gar-On Yeh

16 Rural Governance
International perspectives
Edited by Lynda Cheshire, Vaughan Higgins and Geoffrey Lawrence

17 Global Perspectives on Rural Childhood and Youth
Young rural lives
Edited by Ruth Panelli, Samantha Punch and Elsbeth Robson

18 World City Syndrome
Neoliberalism and inequality in Cape Town
David A. McDonald

19 Exploring Post Development
Aram Ziai

20 Family Farms
Harold Brookfield and Helen Parsons

21 China on the Move
Migration, the state, and the household
C. Cindy Fan

22 Participatory Action Research Approaches and Methods
Connecting people, participation and place
Sara Kindon, Rachel Pain and Mike Kesby

23 Time–Space Compression
Historical geographies
Barney Warf

24 Sensing Cities
Monica Degen

25 International Migration and Knowledge
Allan Williams and Vladimir Baláž

26 The Spatial Turn
Interdisciplinary perspectives
Barney Warf and Santa Arias

27 Whose Urban Renaissance?
An international comparison of urban regeneration policies
Libby Porter and Katie Shaw

28 Rethinking Maps
Martin Dodge, Rob Kitchin and Chris Perkins

29 Rural–Urban Dynamics
Livelihoods, mobility and markets in African and Asian Frontiers
Jytte Agergaard, Niels Fold and Katherine V. Gough

30 Spaces of Vernacular Creativity
Rethinking the cultural economy
Tim Edensor, Deborah Leslie, Steve Millington and Norma Rantisi

31 Critical Reflections on Regional Competitiveness
Gillian Bristow

32 Governance and Planning of Mega-City Regions
An international comparative perspective
Jiang Xu and Anthony G.O. Yeh

33 Design Economies and the Changing World Economy
Innovation, production and competitiveness
John Bryson and Grete Rustin

34 The Globalization of Advertising
Agencies, cities and spaces of creativity
James R. Faulconbridge, Jonathan V. Beaverstock, Corinne Nativel and Peter J. Taylor

35 Cities and Low Carbon Transitions
Harriet Bulkeley, Vanesa Castán Broto, Mike Hodson and Simon Marvin

Not yet published:

36 New Economic Spaces in Asian Cities
From industrial restructuring to the cultural turn
Peter W. Daniels, Kong Chong Ho and Thomas A. Hutton

The Globalization of Advertising

Agencies, cities and spaces of creativity

**James R. Faulconbridge,
Jonathan V. Beaverstock,
Corinne Nativel and Peter J. Taylor**

LONDON AND NEW YORK

First published 2011
by Routledge
2 Park Square, Milton Park, Abingdon, Oxon OX14 4RN

Simultaneously published in the USA and Canada
by Routledge
711 Third Avenue, New York, NY10017

Routledge is an imprint of the Taylor & Francis Group, an informa business

Typeset in Times by Wearset Ltd, Boldon, Tyne and Wear

British Library Cataloguing in Publication Data
A catalogue record for this book is available from the British Library

Library of Congress Cataloguing in Publication Data
The globalization of advertising/James R. Faulconbridge ... [*et al.*]. – 1st ed.
p. cm.
Includes bibliographical references and index.
1. Advertising. 2. Globalization–Economic aspects. I. Beaverstock, Jonathan V.
HF5823.G563 2010
659.1′042–dc22 2010027330

ISBN: 978-0-415-56716-9 (hbk)
ISBN: 978-0-203-86089-2 (ebk)

Contents

Figures

Tables

Acknowledgements

This book is derived from project development at Loughborough University (UK) as part of the GaWC (Globalization and World Cities) programme of research. All four authors have been associated with Loughborough in some capacity but 'academic churn' means that we are all now in pastures new. However, we record our appreciation of the opportunities for research that Loughborough University afforded us and thank our colleagues at GaWC for their support. Beyond its origins, a number of other individuals and institutions made this book possible. First and foremost we are grateful to the Alfred P. Sloan Foundation for generously funding the research reported in this book from the project, 'The globalization of the advertising industry: a case study of knowledge workers in worldwide economic restructuring'. In particular, we thank Gail Pesyna at the Foundation for her help, support and encouragement. We also wish to extend our gratitude to all of the advertising executives who took the time to be interviewed, and for some, re-interviewed, in New York, Los Angeles and Detroit, and who generously provided insights into their industry. We would like to thank Gemma Davies for skilfully producing the New York, Los Angeles and Detroit location maps and Andrew Cook for carefully formatting the final version of the manuscript.

James Faulconbridge is grateful for the period of sabbatical leave in 2009/10 granted by Lancaster University which allowed him to draft several chapters for the book. Jon Beaverstock would like to thank his partner, Nicola, and family for allowing him to sacrifice the entire Easter 2010 vacation to complete his contributions to the book. Corinne Nativel is grateful to her partner, Christophe, for his understanding and support during the writing-up period. Similarly Peter Taylor appreciates once again the forbearance of Enid when in 'heavy-work mode' on the text.

Part I

Situating global advertising agencies and cities

1 Introduction

Advertising is the archetypal 'modern' industry. As a key knowledge-intensive business (professional) service, it is innately bound up with processes of contemporary globalization. Indeed, the globalization of the advertising industry has been fuelled by an ever-increasing reliance on advertising to develop, sustain and spread markets for products in a 'global consumer world'. In this book, we develop a new and highly innovative investigation of contemporary trends relating to the advertising industry and its spatial division of labour in globalization. We examine the key actors in processes of globalization in the advertising industry – global agencies – and assess the impacts of their restructuring on the geography of advertising work worldwide and in three US cities: New York, Los Angeles and Detroit. We explore the advertising industry and spatial division of advertising work through a conceptual framework in the first part of the book focused on the firm, cities and restructuring, and then through primary research based empirical analyses of the business of advertising and associated city based labour process in the second and third parts.

Conceptually we draw on three interrelated bodies of theoretical work. First, we use theoretical work on knowledge-intensive business services (also referred to as advanced producer or professional services) to theorize the structure and spatial organization of advertising agencies (for example, Alvesson 2004; Beaverstock 2004; Bryson *et al.* 2004; Daniels 1993; Empson 2001; Faulconbridge 2006; Nachum 1999). The globalization of knowledge-intensive service industries, and in particular accountancy, advertising, architecture, law and management consultancy, has acted as the basis for a number of significant debates in the academic disciplines of economics, geography, management, sociology and others over the past twenty-five years or so. In particular, interest in corporate strategy (Dunning and Norman 1987), the sociology of professionals (Burrage *et al.* 1990; Faulconbridge and Muzio 2007), the management of knowledge workers (Alvesson 2001; Cooper *et al.* 1996) and knowledge generation, capture and exploitation (Empson 2001) have all been developed with explicit reference to the spatial organization of global knowledge-intensive business services firms. In this book, we theorize, through reference to these areas of research, how global advertising agencies manage their office networks and skilled labour to develop advertising campaigns in particular spatial jurisdictions.

Second, we draw on a parallel, but distinct body of work that examines the organization of project work (for example, Engwall 2003; Grabher 2004; Sydow and Staber 2002). This frames our discussion of the multitude of actors involved in developing an advertising campaign and their geographies. Reflecting the now expansive literature on the geographies of cultural industries (Cooke and Lazzeretti 2008a; Power and Scott 2004; Scott 2000) we also examine the role of cities as sites of advertising work but, in doing so, couple our discussion to a third body of work on world or global cities (Sassen 2006a; Taylor 2004). As perhaps one of the significant spatial forms associated with the globalization of services, the role of world or global cities was first highlighted by Hall (1966) and Friedman (1986), and more recently analysed comprehensively by members of the Globalization and World Cities (GaWC) research network (see www.lboro.ac.uk/gawc (accessed 1 September 2010)) including Beaverstock *et al.* (2000) and Taylor (2004). Cities such as London and New York have been shown to be 'command and control points' of the global economy (Sassen 2000) and hubs for the coordination of the activities of business service firms. As such, it is essential to understand the relationship between advertising and world cities to understand the globalization and geographies of advertising in the twenty-first century. In particular, work on global and world cities understands the globalization of economic activities, such as advertising, to be constituted through 'networks' that connect firms' offices and cities together. Flows of capital, knowledge, information and expert labour produce what Castells (2000) calls a 'network society' in which connections between different cities drive the global economy and reproduce the geography of economic activities. We develop this work by showing how forms of interconnectivity and flow facilitate the work of global advertising agencies and reproduce the uneven world city geographies of advertising work and labour processes.

Of course, in doing this we only offer a partial analysis of the advertising industry. Global agencies are but one type of agency with small and medium-sized national agencies being as or, in the eyes of some, more important than global agencies. For example, according to the US Census Bureau in both the New York City and Los Angeles metropolitan areas 86 per cent of advertising and advertising related firms employed less than twenty people in 2004. This is typical of the structure of the advertising industry in most cities worldwide (see for example, Bryson *et al.* 2004 on the structure of service industries in the UK and further discussion of US cities in Chapter 4). In stark contrast, global advertising agencies' offices in major cities like New York and Los Angeles are, according to our research, usually staffed by well in excess of 100 workers and occasionally by more than 1,000 and are, therefore, atypical in terms of the broader industry. However, with the fifty largest global agencies generating in excess of a staggering US$33 billion of worldwide annual advertising revenues in 2008 (*Advertising Age* 2009), they are, in our eyes, worth studying as important organizations in their own right. Global agencies also provide insights into processes of globalization and their operation in and through cities that a study of smaller agencies would be less likely to reveal. Consequently, the rest of the book focuses on global agencies whilst recognizing the limits this creates for developing a broader analysis of the advertising industry.

Methodologies

Our research, funded by the Alfred P. Sloan Foundation (see www.sloan.org (accessed 1 September 2010)), had the grand aim of investigating the twenty-first century geographies of advertising work worldwide and in three different US cities: New York, Los Angeles and Detroit. We employed both quantitative and qualitative methodologies to complete our analysis. The former uses two sources of data: *Advertising Age* Annual Reports from 2000 for information on agency billings; and US Census Bureau and Bureau of Statistics employment data for city employment in 2004 and then at intervals up to the present day. Using SPSS statistical software we have been able analyse the available data and map the changing global distribution of agency billings, holding company profits and city jobs at the beginning of the twenty-first century. This has helped us to sketch out some of the significant advertising industry and employment and working changes over recent years, something we then analyse in depth using our second, qualitative research design.

Our second (and main) methodology involves analysing data collected from a series of in-depth interviews with key stakeholders in advertising agencies between March and June 2007. Interviewees held a variety of important positions, including chief executive at the top of the hierarchy, in strategic roles (e.g. chief trend spotter), and in creative work (e.g. chief creative officer). The Appendix provides a breakdown of the profile of the advertisers interviewed as part of the research. Throughout the book we have removed the identity of interviewees and their employing agency, something agreed with all interviewees to ensure the frankness of discussions. In the rest of the book extracts from interviews are, therefore, presented as anonymized quotations with interviewee number used to allow the reader to understand the profile of the individual being quoted (see the Appendix for the interviewee key). We can, however, reveal that interviewees worked at truly global advertising agencies including BBDO, Euro RSCG, Publicis, Young & Rubicam, Lowe Worldwide, Ogilvy & Mather, J. Walter Thomson, Leo Burnett, TBWA\Chiat\Day and Saatchi & Saatchi. Transcripts from interviews were all coded using the logic of grounded theory (Glaser and Straus 1967) to reveal differentiated spatial divisions of advertising work within global agencies in the twenty-first century.

The reader will have already realized that the interviews upon which the analysis draws were completed before the credit crisis of 2007 and the ensuing recession. We believe, however, that the analytical framework that we develop in the book based on analysis of change in the years leading up to 2008 provides a useful way to make sense of the severity of the effects of the crisis and recession on advertising work in our three different US cities. To evaluate this premise we completed a series of follow-up interviews (see Appendix) with senior agency executives in January 2010. Whilst it is not possible to use the small set of interviews to reliably identify all of the processes of change that have occurred as a result of the credit crisis and recession, the interviews do allow us to assess our premise about the influences on the severity of the impacts of the crisis and recession in New York, Los Angeles and Detroit.

Qualitative data collection in the 2007 and 2010 interviews was carefully targeted in its geographical scope. Interviews were conducted in New York City, Los Angeles and Detroit because these cities offer a range of critical insights relating to processes of advertising globalization that are repeated in many cities worldwide. New York City was chosen, first, as a city associated with the birth of global advertising agencies and, second, in the contemporary period as a leading world city 'commanding and controlling' agency work. Los Angles was chosen as a global cultural capital because of the presence of the motion picture industry and as the US gateway to the Pacific Asian. Detroit was chosen as an 'advertising city' built upon the presence of a major consumer industry: car manufacturing. The three cities were also chosen because of their changing fortunes over recent years as the geography of global advertising work has been restructured. New York City has sustained itself as a leading global advertising city, Los Angeles has developed a new role above and beyond that associated with the development of advertising for the motion picture industry and Detroit has suffered structural decline because of its ties to the US car manufacturing industry. As such, the three case studies offer insights into the experiences of different cities of the changing geographies of advertising work, something which allows us to extrapolate a broader understanding of the causes and effects of change in the spatial division of advertising work worldwide. Despite being based predominantly on a study of the industry in the US, the book provides, therefore, original and in-depth research findings to understand broader processes of advertising globalization and restructuring occurring worldwide.

The main argument

Our research suggests that increasingly, advertising work is occurring not only in the 'traditional' western core cities of New York, London etc., but also in 'consumer cities' worldwide, particularly in the Middle East and Asia, as part of the emergence of international spatial divisions of labour in global advertising agencies. The ever-growing need for advertising to be created 'close' to the consumer to ensure cultural alignment, timeliness and effectiveness in an increasingly advertising saturated world, has resulted in campaigns which are more and more being produced *in situ* for the 'local' market they target. This means cities such as New York and London have a vital role in serving their respective domestic markets and continue to choreograph global campaigns. But, it is no longer the case that one advert is always exported worldwide, from New York or another 'core' western world city. Instead collaboration between agency offices, in developed and developing economies, is being used to create multiple market campaigns, thus generating more, not less, advertising work and a changing geography for this work. As such, in the advertising world changing spatial relations and divisions of labour are not proving to be a pernicious 'zero sum' game with one city losing work to another. Rather the globalization of advertising in the twenty-first century is producing new strategic centres of work that play a complementary role alongside incumbent centres such as New York.

Simultaneously, however, our analysis shows that cities cannot thrive if they exclusively rely on insular, intra-city economic processes. Cities need a combination of territorial assets – 'local' consumer markets in need of servicing and skilled labour pools to provide the services – *and* network assets – agencies that not only serve 'local' markets, but also play a strategic role in worldwide campaigns – if they are to sustain their role as a strategic site of economic (advertising) activity. As such, a city needs to hold a strategic position in spatial divisions of advertising labour, being connected to other cities by network flows of trade, knowledge, information and talent, which generate work and complement the demand emerging from 'local' markets. We, therefore, explore conceptually and empirically the way variations in the strength of territorial and network assets affect the role of different US cities in advertising globalization.

In making this argument we have to deal with a number of issues of terminology. First, advertising agencies have been variously classified as knowledge-intensive business services, producer services and professional services. There is an extensive debate about which terminology is most appropriate (see for example, Alvesson 2001; Von Nordenflycht 2010). Rather than getting into such debates here, we adopt the term knowledge-intensive business service throughout for reasons justified later in the book.

Second, we take economic globalization to be processual and not an end state and this shapes the nature of our analysis throughout the book. As Dicken (2003: 1) argues, 'there are indeed globalizing processes at work in transforming the world economy into what might reasonably be called a new geo-economy'. However, in studying these processes of globalization we face the problem that, amongst others, Dicken (2003) and Held *et al.* (1999) have highlighted in relation to terminological slippage and fuzziness in descriptions of globalization. For us there are two dimensions to this slippage and fuzziness that are particularly problematic. First, in work on knowledge-intensive business services the terms global, international, multinational and transnational often get used interchangeably (on which see, Allen 1995). In different literatures, each term is used to refer to similar processes whereby there is interaction, collaboration and cooperation between the different subsidiaries of a firm operating in multiple countries. However, as the work of Bartlett and Ghoshal (1998) makes clear, global, international, multinational and transnational are actually very different organizational forms involving different degrees and forms of interaction and collaboration. Firms choose to adopt one of the four organizational forms for strategic reasons. Consequently, throughout the book we use the term 'global agency' in a descriptive manner to refer to an agency with offices in all of the major economic regions worldwide. When reviewing the literature on knowledge-intensive business services in Part I of the book we use the language employed by the authors whose work we review, even when slippage is apparent. We feel it would be inappropriate to change the terminologies used in their analyses. But, then in our analysis of empirical material in Parts II and III of the book which refers explicitly to the organizational form of agencies we use Bartlett and Ghoshal's (1998) terminology to recognize the differences between global, international, multinational and transnational organizational forms, and the adoption of the latter over recent years by advertising agencies.

We also face terminological issues in relation to work on global and world cities. Sassen (2006a) suggests that both terms actually allude to the same phenomenon, despite the effects of the different scholarly traditions on authors' analyses of world cities (world systems theory) and global cities (political economy). Consequently, for sake of consistency, we adopt the term world cities in Parts II and III of the book. However, it should be noted that the processes alluded to in reference to world cities are the type of transnational processes that Bartlett and Ghoshal (1998) describe with networks of interaction, flow and collaboration between cities being significant in spurring economic development, restructuring and change. We hope the reader will bear with us as we navigate this terminological minefield throughout the book.

The structure of the book

The book proceeds in three distinctive parts, containing nine further chapters. Part I sets the scene of our analysis by conceptualizing the role and practice of global advertising agencies as knowledge-intensive firms in the twenty-first century. It does this by developing a framework through which we interpret the new and original research findings presented in Parts II and III. Chapter 2 begins by examining the development of the advertising industry and the emergence of global agencies. It then examines the industry and its current form and the implications of economic changes for the structuring of global agencies. The trends are then explored conceptually through theoretical debates concerning knowledge-intensive business services. The discussion reveals how global advertising agencies have adapted to ever-changing consumer audiences and associated geographies of advertising demand and work, resulting in important changes in the type of spatial relationship associated with advertising work in global agencies. Chapter 3 then develops this argument by considering in detail the role of cities in advertising globalization and the way the strategies of agencies are designed to exploit the benefits of operating in world cities. Particular attention is paid to the way processes of learning, project working and user-led innovation occur in *and* through networked cities. The chapter ends by highlighting the main conceptual understanding developed in the first section of the book: that advertising globalization as a process occurs in and through world cities and that it is a combination of the *territorial and network assets* of a city, assets that reciprocally produce one another, that determines the spatial division of advertising labour.

Part II begins with Chapter 4, which uses the quantitative data collected by the authors to unpack the contemporary global geographies and patterns of advertising connectivities, inter-city relations and work. Chapter 5 then develops an empirical analysis of the way work is organized in global agencies and uses this to explain the geographies and patterns of worldwide advertising inter-city change detailed in Chapter 4. The discussion reveals how the multitude of different tasks and skills associated with the production of advertising in global agencies manifests itself in a topological geography of work which is determined both by the importance of presence in particular cities and by the importance of network relations between cities worldwide.

Part III considers how the changing geographies of advertising and agencies, as identified in Parts I and II, have affected three leading US advertising centres and considers their future role in global advertising work. Chapter 6 examines the implications of contemporary globalization for the city widely acclaimed as the birthplace of advertising: New York. The city's changing role, from command and control centre for global campaigns to strategic collaborator, is examined and the implications for the characteristics of the city's advertising industry are considered. New York is shown to be in an advantageous position when it comes to capturing a consumer audience. In addition, when coupled to the fact that New York continues to act as the 'lead' office for many global accounts this means the city's advertising industry remains healthy thanks to strong territorial and network assets.

Chapter 7 examines Los Angeles, as the centre of the global motion picture industry, and the contemporary role of the city in advertising globalization. Using our data we reveal that, surprisingly, the cooperation between advertising agencies and the entertainment/cinema industry appears to be weaker than expected, the underlying reason being that filmmakers adopt a different approach from other industries to reach consumers, using specialist in-house or external media companies. Instead our data shows that automobile clients feature very strongly within LA agencies' portfolios. However, LA demarcates itself strongly from Detroit as these clients are Japanese car manufacturers such as Toyota or Nissan as well as the US giants such as Ford. The Californian lifestyle is shown to be an important reason for LA having such a vibrant industry, something attractive for brands which are regarded as stylish and, in terms of cars, fuel efficient and environmentally friendly, something not associated with Detroit. Indeed, one of the main ideas put forward in this chapter is that Los Angeles is less 'global' than might be expected, its key role being as the 'capital' of 'west coast' alongside its role as gateway from/to Pacific Asian. As a result, the city has gained from the increasing decentralization of advertising control from New York as 'cultural proximity' to west coast consumers is sought. Again we interpret these findings in the context of understandings of the organization of advertising agencies and relational network connectivities in the global economy developed in previous chapters. This explains the role of the city in advertising globalization as being tied to both territorial and network assets of variable strength.

Chapter 8 examines Detroit, the car city, as a city that has always had a major advertising industry serving US car manufacturers. With the decline of those manufacturers, competition from overseas producers and the downwards spiral Detroit seems to have entered, we show how major questions exist about the city's advertising industry. The chapter shows how Detroit agencies may have had an advantage in the past because of the proximity of key clients (territorial assets in terms of 'local' demand). But, today this advantage is waning since these clients undergo restructuring themselves as their markets come under threat and begin to look to offices and agencies in other cities within and without the US for advertising. The chapter also shows that a related problem for Detroit agencies exists because, in the world of advertising, the creative work performed in Detroit is perceived to be somewhat stale and backward. The most enterprising younger talent is less likely to stay in the city. We interpret and highlight the implications of these trends through work on world cities and their creative clusters and explain the repositioning of Detroit with reference to earlier discussions of changing organizational forms in global advertising agencies. Specifically we suggest Detroit has increasingly weak territorial and network assets.

Chapter 9 considers the implications of the previous chapters' discussions of the role of New York City, Los Angeles and Detroit in advertising globalization in light of the impacts on the advertising industry of the credit crisis of 2007 and ensuing recession. We explore the impacts on each of the three cities and relate them to discussions of the strength and weakness of the territorial and network assets. This reveals the usefulness of our analytical framework for studying the resilience and changing strategic role of different cities in economic globalization over time.

The Conclusion draws the discussions in the preceding chapters together and reflects back on the theoretical significance of the findings presented throughout the book. The policy implications of the book's findings are also identified in this chapter. Theoretically, we highlight the way connecting work on professional services to work on project work and studies of advertising reveals new insights into the organization of work in global firms. We also reveal how relational/network approaches to studying global firms and cities can be effectively used to develop understanding of the effects of the changing geographies of the global economy on incumbent cities (such as New York and Los Angeles) and the broader changing geographies of knowledge-intensive work in the global economy. An agenda for future research is proposed based on these highly original and innovative research findings. In terms of policy, we argue that the future of the advertising industry in US cities (and other cities worldwide) depends on the ability both of industry leaders and policy makers to recognize the complex, multidimensional and city-specific opportunities and threats faced rather than adopting a one dimensional one size fits all approach to policy.

2 The global advertising agency

The globalization of advertising has its origins in the early part of the twentieth century. American agencies began the process of opening overseas offices prior to the two World Wars and accelerated their globalization throughout the latter part of the twentieth century. For example, the agency McCann Erickson, which was established in New York City in 1902, opened its first European offices in 1927, followed by offices in Latin America in 1935 and Australia in 1959. Such globalization strategies were tied to specific clients, in particular to the emergence of a cadre of manufacturing transnational corporations (TNCs) in the early to mid twentieth century (see for example, Dicken 2007; Thrift and Taylor 1989). McCann Erickson followed its main client – Standard Oil – as it globalized, whilst other agencies such as J. Walter Thomson – key client General Motors – adopted a similar strategy in order to provide advertising services wherever clients operated (see for example, Perry 1990; West 1987). Latterly, in the 1960s and 1970s, English agencies also began to recognize the overseas opportunities associated with globalization (Leslie 1995). For example, Saatchi & Saatchi, perhaps one of the most iconic English agencies, was founded in 1970 and quickly developed a global network of offices on the back of relationships with clients such as British Airways and Toyota. More recently a number of Japanese agencies have also followed suit, most notably Dentsu, initially through its now ended collaboration with the US agency, Young & Rubicam.

It is not our intention here to further explore the history of individual global advertising agencies or the advertising industry more broadly. This has been done by others (see for example, Alter 1994; Clarke and Bradford 1989; Fendley 1995; Lears 1995; Mattelart 1991; Perry 1990; Roberts 2004). Instead, we discuss the relationship between the geography of global advertising agencies, cities and the geography of advertising project work. First, we examine the role of global advertising agencies within the advertising industry more broadly in the early twenty-first century. Second, we consider how the structures and strategies of global advertising agencies in the twenty-first century can be understood in the context of academic work on knowledge-intensive business services. This reveals the way the book's three main themes – the spatiality of markets, work and innovation – are understood by existing literatures on knowledge-intensive business services and provides a framework to interpret the original empirical research findings relating to the organizational strategies of contemporary global advertising agencies.

The main argument in this chapter is as follows: the spatial strategies of global advertising agencies in the early twenty-first century can be explained through analysis of the intersecting need to serve geographically heterogeneous markets and engage in project work that is equally reliant on territorial (local) and network (global) forms of innovation and knowledge. By considering the competing, but interrelated demands of the 'local' and the 'global', we contend, it is possible to understand the contemporary geographical strategies of global advertising agencies in the world space economy.

In analysing the work of global advertising agencies we are focusing principally on their activities that involve serving client TNCs and managing global accounts which require advertising campaigns to be run in multiple, geographical markets. This is not the only work global agencies do. They also serve clients that require campaigns in single countries or regions. However, because of our interest in the relationship between global agencies, the globalization of advertising and the geography of advertising work in the twenty-first century, it is instructive to explore the way such global accounts are handled because they provide a lens through which we can understand the grounding of global advertising work, and client relations, in particular cities.

Global agencies and the advertising industry

As a product of the twentieth century, the global advertising agency exists as one component of a more complex advertising industry that has come to be one of the central 'lubricators' (Dicken 2007) or 'fixers' (Thrift 1987) of contemporary capitalism. In order to understand the current role of global advertising agencies it is, therefore, important to first understand the role of the advertising industry in particular capitalist regimes of accumulation. As a tool to avoid a mismatch between production and consumption, advertising came of age during what has been called the Fordist era of mass production (see for example Harvey 1989a). Advertising acted as a means to stimulate the mass consumption needed to support mass production systems and, as such, a whole new knowledge-intensive business service industry was born in the first half of the twentieth century (see for example, Leslie 1997a, 1997b; Schoenberger 1988). During the Fordist era, advertising as a service industry principally existed as a tool to 'educate' consumers about the availability of a product. Referred to as first wave advertising, campaigns focused upon description and fact to sell products. Such adverts were relatively easy to produce based on quantitative analysis of generic target audiences and their lifestyles which were classified into a limited number of demographic segments (Marcuse 1964). As a result, global advertising in this period effectively operated as a US export industry. Adverts for US global clients, such as Coca-Cola®, were produced in the USA with a set of universal principles and strategies developed to design campaigns that ran worldwide (de Mooij and Keegan 1991; Mattelart 1991).

In the latter part of the twentieth century, advertising took on a new role in the context of what can be crudely classified as a post-Fordist era. For Lash and Urry (1994: 123) advertising as a cultural industry is, 'post-Fordist avant la lettre', in that the emergence of newly reflexive consumers in the later part of the twentieth century created both new challenges but also new demand for advertising (Crang 1996; Crewe and Lowe, 1995). Lash and Urry (1994: 111–12) suggested that the post-Fordist era witnessed the development of consumers that were both 'cognitively' and 'aesthetically reflexive', something which Miller (1995: 48) suggested meant advertisers needed to be in collusion with the consumer in order to develop effective advertising. Such collusion – a key feature of what has become known as 'second wave advertising' – required more and more intimate knowledge of smaller and smaller market groups as advertising targeted at macro scale categories, such as 'single women', and designed to stimulate a homogeneous form of mass consumption, became redundant. In its place new advertising emerged designed to be reflexive and able to respond to the intersecting identities and complex positionalities of consumers that value individuality over mass consumerism. This was quickly followed by so-called 'third wave advertising', which involved tailoring adverts not only to specific reflexive consumers, but also to their reactions and interpretations of recent global- or country- or region-specific events. Adverts began to appear just weeks after key political or sporting events and attempted to reflect consumers' responses to them.

As Leslie (1997b) outlines, evolutions associated with 'second' and 'third wave advertising' led to upheaval in the advertising industry. Although demand for the services of advertisers grew and thus created new business opportunities, the advertising required by agencies' clients was more and more sophisticated, and relied upon more and more creative campaigns. For global agencies, initially born to serve the need of manufacturing TNCs to stimulate mass consumption wherever they operated, this meant responding to the new desire of their clients to stimulate demand from reflexive consumers in multiple markets worldwide. As a result, the export led model of the Fordist era faltered in a post-Fordist epoch characterized by increased heterogeneity in consumer identities both between consumer groups and across space. Particularly significant in relation to our argument here is the fact that, as part of the emergence of post-Fordist regimes, global agencies had to develop the ability to create demand for a product in multiple geographically dispersed and heterogeneous markets in which consumer behaviours and relationships to products differed.

There is now an extensive literature that outlines the nature of such geographically heterogeneous consumers and their place-specific relationships to products (see for example, Lury 2004; Molotch 2003a; Pike 2009; Weller 2008). Here we summarize the key findings of this research and its implications for the work of global advertising agencies. It has been shown that the relationship between consumers and particular brands and products is defined by a complex array of spatial entanglements. Pike (2009) develops a sophisticated argument in this regard and shows how consumers identify with brands or products based on two types of spatial entanglement. First, brands or products can themselves be associated with particular places. For example, Pike draws on the work of Castree (2001) to highlight how products such as Uncle Ben's rice or the Jeep Cherokee have, through their portrayal in advertising, become associated with particular places and cultures. Second, brands and products and consumers' relationships with them are produced by spatially situated interpretations of any advertising. The way consumers relate to a brand or product, and react to advertising promoting it, is determined by the way, 'language, symbols, colours and consumer preferences remain heterogeneous and geographically differentiated' (Pike 2009: 635). The work of Dwyer and Jackson (2003) and Dwyer and Crang (2002) shows, for example, that products as diverse as women's clothing and curry sauces have place-specific meanings that are tied both to the way consumers understand the geographical origins of a product and its role in their life as determined by a range of place-specific entanglements, which create situated understandings of fashion and diet, for example.

At one level, such spatially heterogeneous consumers have created a challenge for the producers of consumer goods themselves. The inability of car manufacturers to produce a global car is a well documented example of this with region-specific models being needed to respond to different consumer tastes (Dicken 2007). Molotch (2003a, 2003b) describes how such diversity also exists within countries with Detroit based car manufacturers setting up design studios in California in the 1970s to allow cars suited to the very different west coast market in the USA to be developed. At another level, such spatial heterogeneity also creates major challenges for advertisers. Place-specific product variations tend to prohibit global adverts and even global products, such as Coca-Cola®, have developed place-specific identities as situated social practices, influenced by what Molotch (2003a: 672) calls 'semiotic handles', create spatially contingent consumer–product relationships. For Weller (2008) this means the stabilization of a product's or brand's identity is always a situated social construction, entangled with events, activities, images and the way individuals and groups make sense of the product or brand in question. Meanwhile for Aspers (2010: 8), those targeting their products or adverts at spatially entangled consumers have to be part of the 'lifeworld' of the consumer and understand 'what people take for granted and do not question, such as basic values, propositions, facts, culture and so on'. This means being co-located with the consumer so as to develop contextual knowledge of their 'lifeworlds'.

As a result, in the late twentieth and early twenty-first centuries, only by understanding the various spatial entanglements and situated influences on the way consumers use, engage with and react to any product and its associated advertising, is it possible to develop effective advertising campaigns. As documented by Leslie (1997b), global agencies have, therefore, increasingly found themselves in competition with smaller single office 'boutique' agencies because of their ability to understand and respond to 'local' situated consumer identities and product relationships. These smaller agencies, whilst tending to serve clients seeking to stimulate demand for their products in one market only and often only employing tens of workers in one location in contrast to the tens of thousands employed worldwide by global agencies, have been highlighted as the winners of the post-Fordist advertising era because of the advantages of being located in one place and serving the *in situ* 'local' market. Thus global producers of consumer goods, including Coca-Cola®, began experimenting with using multiple small boutiques to develop campaigns for different markets throughout the world, rather than relying on one agency and one global campaign.

As a result, in the late twentieth century global advertising agencies had to prove that they were capable of fulfilling the demand of clients for advertising in multiple markets that responded to the spatial heterogeneity of consumers and the situated entanglements affecting relationships with products and brands. The rest of the chapter deals with the way global agencies, through their changing contemporary organizational structure, have developed such responsiveness as part of a shift away from the export model of what Mattelart (1991) calls 'imperial' advertising and the effects of this shift on the geography of advertising work.

Global advertising agencies within holding group structures

Deciphering the 'world' of global advertising agencies in the twenty-first century is no simple task. In part, this is because of the *holding group structure* now widely adopted within the industry, and the way *agencies* operate under these umbrella structures. Next we explore the characteristics of holding groups before focusing, in detail, on the agencies and their advertising work.

Holding groups

Holding groups act as financial conglomerates that absorb all of the profits and costs of agencies within the group and, as such, are the stock market listed entities in which shareholders can invest. Since the early 1980s in particular, the advertising industry has been undergoing rapid consolidation through a continual merger and acquisition process which has produced the large holding groups listed in Table 2.1. For example, the agency Saatchi & Saatchi transformed itself into a holding group in the 1980s by purchasing agencies including Ted Bates and Backer & Spielvogel, before the Saatchi & Saatchi group itself was purchased by Publicis in 2000/01. This merger and consolidation process has continued in the twenty-first century and has been accelerated by periods of recession associated with the dot.com bust of 2000. For example, since 2000 other major mergers have included the Cordiant Communications group being acquired by the WPP group in 2003, resulting in the closure of the Bates agency, and in 2005 the Grey Global group of companies also joining the WPP group.

The promotional literature of WPP captures the supposed client benefits of such holding groups. As WPP proclaims:

> Through our companies and associates, WPP offers a comprehensive and, when appropriate, integrated range of communications services to national,

Table 2.1 The key global communication groups and their advertising agencies

Communications group	*Revenues (billions US$, 2009)*	*Global employees (2008)*	*Main global agencies*
Dentsu	3.1	N/A	Dentsu Incorporated
Havas	2.0	14,747	Euro RSCG Worldwide; Arnold Worldwide
Interpublic	6.0	43,000	Draft Foote Cone and Belding Worldwide; Lowe Worldwide; McCann Erickson Worldwide
Omnicom	11.7	70,000	BBDO Worldwide; DDB Worldwide Communications; TBWA Worldwide
Publicis	6.3	43,808	Leo Burnett Worldwide; Publicis Worldwide; Saatchi & Saatchi
WPP	13.6	90,182	J. Walter Thomson; Ogilvy and Mather Worldwide; Young & Rubicam; Grey

Source: *Advertising Age* 2009, 2010a.

> multinational and global clients. WPP companies work with over 340 of the Fortune Global 500; over one half of the NASDAQ 100 and over 30 of the Fortune e-50. Over 600 clients are served in three disciplines. More than 370 clients are served in four disciplines; these client account for over 58% of Group revenues. The Group also works with over 270 clients in six or more countries.
>
> (2008: 1)

In reality, however, the main motivation for the creation of holding groups is the opportunities for revenue growth. Since the early days of the advertising industry and in response to client demands, agencies have only worked for one client in any one industry. Agencies have, therefore, been founded on long-term client relationships, such as those between McCann Erickson and Standard Oil in the early twentieth century and more recently Ogilvy and Mather and IBM. Concern about insider knowledge reaching competitors and the inherent conflict of interest involved in developing campaigns for competing clients (for example, Ford and General Motors) lies behind the existence of such exclusive relationships, thus, limiting the ability of agencies to grow because of the inherent restrictions associated with working for only one client per industry. Of course, it does not make commercial sense to have to choose between clients and lose a lucrative contract because of conflict of interest concerns. Thus the umbrella of the holding group was born to allow 'Chinese walls' to be created between agencies whereby clients in competition can be serviced within the same group but by different agencies, thus allowing additional growth of revenues and profits. This is the fundamental rationale of the twenty-first century holding group, with surprisingly limited interaction occurring between agencies within groups as a result.

The holding group–agency distinction also corresponds to a division of labour: the holding companies do the financial management and accounting associated with operating as a global business, and the advertising agencies design and deliver the product and creative media services. Therefore, it is not the holding groups that are of interest in this book. Instead it is the agencies and, more specifically, the workforce and projects within agencies and the enactment of global advertising work.

Agencies

In their current guise, the leading global agencies (Table 2.2) have grown within their holding groups to become, alongside accountants (Beaverstock 1996) and management consultants (Jones 2003), perhaps some of the most geographically dispersed knowledge-intensive business service organizations (see for example, Beaverstock *et al.* 1999a and Faulconbridge 2008 on the relatively limited reach of 'global' law firms in comparison to advertising agencies). As *Advertising Age* (2000) noted, 'the global land grab' by agencies is now almost complete with all of the ten largest global agencies having offices on every continent including Africa.

The 'land grab' described by *Advertising Age* (2000) formed a central plank of agency strategies throughout the 1990s (Leslie 1995) and into the 2000s (Faulconbridge 2006). As such, global agencies have an impressive geographical reach in the form of an *in situ* office presence in multiple markets throughout the world, something that acts as a central part of the corporate identity and unique

Table 2.2 The ten leading global agencies by revenue

Company	*Worldwide revenue (billions in 2009)*	*Global offices*
McCann-Erickson	$1.41	417
BBDO worldwide	$1.14	287
DDB Worldwide	$1.11	200
J. Walter Thompson	$1.07	227
TBWA Worldwide	$1.02	233
Young & Rubicam	$0.93	304
Euro RSCG Worldwide	$0.93	84
Draft/FCB	$0.89	106
Publicis Worldwide	$0.88	170
Saatchi & Saatchi	$0.65	150

Source: *Advertising Age* (2010a) and authors' research.

service offering promoted by global agencies. For example, Ogilvy & Mather celebrates on its website the fact that:

> Ogilvy & Mather is one of the largest marketing communications networks in the world, with 450+ offices in 120 countries ... The quality of our network is based on the strength of our international network, our local strength and depth across all communication disciplines, our culture of collaboration and our people.
>
> (www.ogilvy.com/About/Network/Ogilvy-Mather.aspx (accessed 1 September 2010))

In the rest of this chapter we examine a range of important questions about the role of the numerous offices of global advertising agencies, the connections between them and the way the organizational structures developed in the late twentieth and early twenty-first centuries allow market-specific advertising to be developed that responds to reflexive consumers influenced by complex spatial entanglements. We do this by situating our analysis in relation to existing work on knowledge-intensive business services. In doing so we begin to develop a conceptual framework that can be used in the rest of the book to empirically interrogate the geographies of advertising work, something we do in Part III through reference to our three case study cities: New York, Los Angeles and Detroit.

Global advertising agencies as situated and interconnected knowledge-intensive business services

Perhaps one of the most important ways of understanding the global advertising agency is as an externalized knowledge-intensive business service. There are many debates about which industries actually classify as knowledge-intensive (see for example, Alvesson 2004) and we justify our use of this term as follows. In terms of the use 'business service', advertising agencies, as providers of tools for dealing with the contemporary complexities of aligning production and consumption, are a form of externalized business service with clients relying on independent agencies, rather than in-house departments, for the production of campaigns crucial for generating consumer demand. As such, advertising is not a consumer service, despite the fact that all work is inherently tied to interactions with consumers through campaigns. Advertising does warrant the label, 'business service', i.e. a service purchased exclusively by businesses.

In terms of the use of the term 'knowledge-intensive', our justification is associated with the way that advertising agencies provide access to knowledge and expertise that is not part of the client firm's core competencies. As such, advertisers provide the type of 'bespoke' problem solving services that Empson (2001) identified as a distinguishing feature of all knowledge-intensive business services, with the knowledge-base of the agencies being associated with both industry-specific expertise but also client service and relationship management.

Table 2.3 The key factors for success in advertising

Key factor	*Reason for importance*
Creativity	'add[s] value by manipulating existing knowledge' (p. 31).
Quality of employees	'The ability of employees to manipulate existing knowledge in order to solve specific client problems' (p. 33).
Reputation	'The intangible nature of the output of these firms creates difficulty to assess its value . . . extensive human involvement in the production and the great need for tailor-made solutions [means] the "same" service might differ every time' (p. 35).
Client–supplier relations	'The interaction between the service provider and the client is considered to be one of the unique features of services' (p. 36).
Age	'Age is typically associated with accumulated experience' (p. 38).
Size	'Large firms can gain economies from greater specialisation of professionals' (p. 39).
Economies of scope	'Such economies take place across segments, products or markets and may involve the joint use of different kinds of assets' (p. 41).
Organizational structure	'The need for an in-depth knowledge of the domestic market limits the ability of headquarters based in another country to guide and control work of affiliates based elsewhere' (p. 43).
Managerial capabilities	'Managers have to deal with high-quality and high-cost labour . . . the need to produce creative work requires special managerial efforts' (p. 44).

Source: adapted from Nachum 1999: 31–45.

Nachum (1999) develops this idea and lists nine key elements of the expertise that successful advertising agencies possess and which clients seek to access (Table 2.3). Each factor in some way reflects the importance of knowledge and expertise in the acquiring, designing and executing of advertising work. As such, the core competency of an advertising agency is, 'knowledge of how to meet specific needs of clients' (Nachum 1999: 31), and knowledge which is possessed by the agencies' workers in the form of what Polanyi (1967) might refer to as tacit knowledge or what Blackler *et al.* (1998) would call a mix of encultured knowledge – knowledge that individuals develop and which allows sense making in relation to the problem presented by a client – and embrained knowledge – knowledge that allows entrepreneurial problem solving.

Consequently, reliance on the tacit knowledge of executives is one of the defining features of advertising agencies. This is special knowledge that allows market research to be judged, interpreted and then applied in a tailored, unique way so as to fulfil a client's needs. As Hackley (1999: 721–22) argues:

> The highest levels of expertise in this domain, as in other professional domains, depend upon an interaction of domain relevant knowledge, experience and creativity ... Large amounts of marketing knowledge are codified in popular texts and constitute a public discourse. However, much of the knowledge underpinning practical marketing expertise may be tacit, implicit in the day-to-day problem solving of strategic marketing practitioners.

The major challenge faced by agencies and their clients is, then, that:

> [m]arketing problems ... tend to be 'ill structured' ... and as such require structuring by the problem solver and it is therefore vital that agencies can rely on their workers 'forming and refining a heuristic or rule of thumb for solving their problem. The power of this heuristic in solving the problem depends on the high level of skill of the marketer ... and is founded on extensive knowledge and experience'.
>
> (Hackley 1999: 727)

Our aim is to develop a conceptual framework though which it is possible to understand how global advertising agencies, as knowledge-intensive business services, have used globalization and the development of worldwide office networks as a strategy to fulfil clients' needs for advertising that responds to reflexive, geographically heterogeneous and entangled consumer behaviours and identities. A useful starting point for this analysis is revisiting the theoretical discourses of globalization and its application to knowledge-intensive business service firms.

Theorizing the 'why' and 'how' of advertising agency globalization

Existing literature attempts to explain the globalization of knowledge-intensive business service industries such as advertising using Dunning and Norman's (1987) well known 'eclectic paradigm'. The 'eclectic paradigm' highlights three core considerations in relation to the 'why globalize' question. First, are ownership advantages. A firm might globalize when it possesses unique firm-specific assets that can be exploited in new overseas markets. Second, are location advantages. Globalization is viewed as appropriate when new place-specific assets can only be exploited by the existence of a subsidiary overseas. Third, are internalization advantages. Firms benefit from globalization when *in situ* presence through an owned subsidiary allows risks associated with sub-contracting production to an overseas firm to be avoided (i.e. it protects global brand integrity).

The three strands of the 'eclectic paradigm' have been used to interpret the globalization of, amongst others, accountancy (Bagchi-Sen and Sen 1997), law (Beaverstock *et al.* 1999a), executive search (Faulconbridge *et al.* 2008) as well as advertising (Bagchi-Sen and Sen 1997; Daniels 1993) knowledge-intensive business services. In the case of advertising it is the relationship of agencies with their clients that receives most attention in such explanations. As noted at the start of the chapter, global agencies emerged out of strategies designed to follow home-country clients overseas and Bagchi-Sen and Sen (1997) suggest that this was the main ownership advantage of firms that drove them to globalize. Consequently, location advantage according to Bagchi-Sen and Sen (1997) relates to the benefit accrued from being in proximity to clients' overseas operations whilst internalization advantages relate to the value of acting as a client's sole provider of advertising worldwide and the cultivation of the long-term business relationship.

From this perspective, the globalization of advertising is very much tied to the globalization of the producers of consumer goods and involves opening overseas, often wholly owned offices, which provide services to clients, *in situ*. However, there are a number of other ways of further explaining the 'why' and 'how' of agency globalization and, in particular, the location advantages accrued through globalization. To uncover these explanations requires, however, the use of a more sophisticated theoretical framing of the globalization of knowledge-intensive business services. Relational (Boggs and Rantisi 2003; Yeung 2005) and global production networks (GPN) (Dicken *et al.* 2001; Coe *et al.* 2008) approaches provide just such a framing, being able to tease out the finer details of agency globalization strategies.

Developed initially through studies of manufacturing TNCs, relational and GPN approaches seek to understand the intimate relationship between firms and places. In particular, GPN approaches seek to develop an understanding of the various ways that the operations of TNCs are embedded by a range of social network relations (Dicken *et al.* 2001; Coe *et al.* 2008). Drawing on the work of Polanyi (1944) and more recently Granovetter (1985), the GPN approach has shown that in order to understand the logic of internationalization and the benefits it brings to firms, it is essential to consider the synergistic effect of multiple forms and spaces of embeddedness that facilitate the production process. Hess (2004) offers a threefold conceptualization of this embeddedness. First, Hess (2004: 177) suggests that firms are territorially embedded and '"anchored" in particular territories or places', not least by a local network of suppliers relied upon in the production process. Second, Hess (2004: 176) highlights the role of societal embeddedness, 'the societal (i.e., cultural, political, etc.) background or – to use a "biologistic" metaphor – "genetic code", influencing and shaping the action of individuals and collective actors'. Finally, Hess (2004: 177) draws attention to the role of network embeddedness, 'the structure of relationships among a set of individuals and organizations'. This focus on embeddedness in the GPN approach allows three further rationales to explain advertising agencies opening new offices in different geographical markets.

First, in relation to territorial embeddedness, the importance of having an *in situ* presence in markets throughout the world can be explained in relation to the need for the management of both the 'supplier' relationships associated with advertising (set designers, printers, acting agencies etc.), and for access to labour markets that provide the skilled executives that design and deliver campaigns. In addition regulatory hurdles, such as the need to produce locally (i.e. film, photograph etc.) all adverts screened in a particular country (as is the case in Australia) or peculiar media regulation regarding content or who can purchase advertising space, can both 'obligate' agencies to have an *in situ* presence (Liu and Dicken 2006) and render an office the only way to serve the market. Second, in relation to societal embeddedness, the need for overseas offices can be understood in relation to their role in the development of knowledge and understanding of situated, geographically variable consumer relationships with a client's products. Third, in terms of network embeddedness, the value of global agencies' office networks can be rationalized with reference to benefits that intra-firm relationships between agency workers in different geographically dispersed offices bring in terms of the facilitation of transnational project management.

We next examine in more detail the issues associated with societal and network embeddedness, which brings work on knowledge-intensive business services into dialogue with work on world cities and projects, reserving discussions of territorial embeddedness for the next chapter. Together, all these arguments help explain the 'why' and 'how' of globalization for the office networks of advertising agencies.

Embedded office networks and the logic of internationalization

Societal embeddedness allows the GPN approach to take account of growing recognition, in light of what has been described as the 'cultural turn' in economic geography (Thrift and Olds 1996), that all economic activities are affected by the social and cultural foundations of production and consumption. For advertising, as a cultural industry that seeks to 'manage' and respond to consumption behaviours, societal embeddedness is especially significant because of the growing complexity of the post-Fordist cultures of consumption. As such, globalization and the establishment of overseas offices has moved on from being a strategy designed to allow the exporting of advertising and the retaining of a presence close to clients' overseas operations. It has become a strategy for also managing the growing complexity of geographically variable consumer cultures and their impacts on the effectiveness of advertising. At its simplest this might mean that overseas offices adapt campaigns produced elsewhere to suit the needs of 'local' markets, for example by refilming a television advert using 'local' actors against a backdrop and wearing clothing that reflect the 'local' market. Such an approach is, however, the most simplistic way of dealing with forms of consumer societal embeddedness and least common. Indeed, we refer to 'local' in inverted commas above because of the way consumer cultures disregard the boundaries of hierarchical spatial scales (i.e. local, regional, national, global) and exist as complex social constructions that are geographically variable and contingent, potentially multi-scalar and best considered outside the constraints of scale defined thinking (see for example, Marston *et al.* 2005).

As such, global advertising agencies have to deal with groups of consumers that have different identities and behaviours that display variability, not simply at the national scale, but at sub-national scales and at spatial sites hard to capture using scalar terminologies. Indeed, sub-national groupings might share some commonalities with groupings in other regions or countries because of transnational connections and commonalities in everyday practices and norms, hence the need in the strictest academic sense to question the use of scalar conceptualizations of advertising markets (Mattelart 1991). We return to this issue again in Chapter 5 when we explore just how agencies categorize the spatiality of consumers and the way this influences advertising work.

As a result of the need for advertising that responds to geographically variable forms of consumer individuality, intersecting identity and spatial entanglement, the office networks of global agencies have taken on a more important role, as an agency's network of offices are charged with designing advertising to deal with the peculiarities of one situated group of consumers. This strategy is a shift away from a few global offices designing global campaigns with most offices simply acting as a 'post box' into which a global advert can be delivered. Overseas offices now provide a way of dealing with the ever more challenging societal embeddedness affecting 'third wave' advertising work. Each office acts as a window into the cultural, economic, political and social world of situated consumers and allows reflexive advertising campaigns to be developed.

It is, however, important to examine the 'new' role of overseas offices – as tools to cope with the societal embeddedness of global advertising – in the context of the broader operations of global advertising firms and the network embeddedness of these operations. The activities of global agencies and their hundreds of offices worldwide are, in important ways, influenced by network embeddedness that exists in the shape of inter-office relationships that allow transnational collaboration and teamwork to be used to develop campaigns (Faulconbridge 2006; Kotabe and Helsen 2001). Such collaboration and teamwork is important not least because it allows TNC clients' desires to develop global brands to be fulfilled. It is, therefore, important to distinguish between brands and adverts at this point.

Brand refers to the name or trademark of a company, such as Coca-Cola®, and the image and identity associated with the company's products (see for example, Lury 2004). Hence branding as a process gives meaning to products and services (Keller 1993; McCracken 1993) and in so doing helps in the process of creating demand and distinguishing a product from its competitors (Aaker and Jacobson 2001). Brands can, therefore, be leveraged to increase shareholder value and market share (see for example Balmer and Greyser 2003; Madden *et al.* 2006; Pruzan 2001). Brand*ing* is, then, used here to refer to a management process and practice designed to devise, stabilize and reproduce a brand, most notably through the development of advertising (Arvidsson 2006). Global agencies provide both consultancy advice about how to best develop a firm's brand, and services relating to the development of advertising campaigns to promote the brand. It is the latter that is our primary concern here.

When a TNC seeks to promote a coherent brand worldwide through advertising, the inter-office networks of global agencies become vital. Agencies will coordinate work across multiple offices to ensure the firm's brand is not compromised or portrayed in a contradictory way by adverts developed by each office running a campaign for the client. Hence global agencies allow global management of brand image and identity so as to maintain consistency, but alongside 'local' management of the portrayal of that brand through geographically tailored adverts that are relevant to a situated audience and which respond to the spatial entanglements that give meaning to a brand in any one place. This requires a fine balance between what is often referred to as worldwide brand 'stewardship' or 'guardianship' and situated advertising that recognizes the subtly different and geographically situated meanings, uses and significance of a product in everyday life. The network embeddedness of global agencies is, thus, one of their unique competitive advantages.

Drawing on the work of Bartlett and Ghoshal (1998) and their typology of organizational forms, de Mooij and Keegan (1991) highlight how such network embeddedness became the hallmark of the transnationally integrated agency in the late twentieth century (see for example, Englis 1994; de Mooij 2004, 2005; Papavassiliou and Stathakopoulos 1997; Sirisagul 2000). Table 2.4 describes various business models used by agencies with the movement from first to third wave advertising corresponding with a movement from international to global and then transnational business models. So, as Snyder *et al.* (1991) note, in the 1960s agencies tended to use the 'global' model and often ran the same advert in

Table 2.4 Four models of agency organization

Agency structure	*Key characteristics*	*Strengths/weaknesses*
International	Engages in business overseas primarily using the resources already available in home-country offices.	Small overseas offices but little or no local knowledge used in advert development; products adapted as seen fit by headquarters; local offices implement adverts as instructed.
Multinational	Independent branches operating in each country developing own strategies and products.	Greatest weakness is inability to see similarities and learn from others work; high level of local knowledge and tailoring of campaigns.
Global	Centralized operations controlling activities for all countries.	May operate without any overseas offices; good at highlighting similarities between nations but applies 'one size fits all' with little concern for local variations.
Transnational	Integrated resources serving clients through development and diffusion of knowledge worldwide.	Responsiveness to local whilst also providing global integration and exchange of learning and ideas; strong strategy outcomes for client balancing global–local tensions.

Source: adapted from De Mooij and Keegan 1991: 8–10.

North America and Europe as part of cross border campaigns. By the 1970s this had evolved so that a more 'international' strategy was in place, with the same advert being used in North America and Europe, but with local adaptations made. The transnational model, which we describe in more detail next, was developed in the late 1980s and 1990s, and allows agencies to overcome the dilemma of being more 'global' than consumers by offering TNC clients the ability to develop some degree of global consistency in brand management, whilst also developing multiple adverts tailored to particular situated markets.

Innovation and management through transnational organizational forms

Global agencies rely on transnational organizational forms to allow the collaborative development of innovative ideas for campaigns. Reflecting the suggestion that the transnational form is designed to replace the 'imperial' export model of the past and respond to spatially heterogeneous and reflexive consumers through market-specific campaigns, this collaboration does not involve the transfer of knowledge between offices. Because of the tacit, encultured and embrained nature of advertising knowledge, the transnational model is used to facilitate a social process of learning through which individuals share ideas, experiences and understandings, thus leading to new ideas emerging and individuals developing new forms of tacit, encultured and embrained knowledge. As Grein and Ducoffe (1998: 312) argue, the transnational form, and the network embeddedness it draws on, 'helps build working relationships, gather information needed for soliciting new international business, solve problems on international accounts, control or evaluate performance, and ensure that the best ideas in the network are being shared'.

Transnational embeddedness and collaboration is important for global agencies because the tacit, encultured and embrained knowledges needed to develop effective campaigns correspond in the schema of Asheim *et al.* (2007) with synthetic and symbolic knowledge forms. Table 2.5 sets out the schema of Asheim *et al.* (2007) in more detail. This schema attempts to conceptualize the way different forms of knowledge diffuse or get shared across space, as determined by the core characteristics of the knowledge-base. As the schema shows, both synthetic and symbolic knowledge are considered hard to transfer and diffuse, requiring either regular face-to-face contact in order to allow individuals to learn from one another, or existing in such a form that is impossible to 'share' and can only be exploited by allowing individuals with the required knowledge to work directly on a project.

The role of the transnational agency form in innovation is to deal with the difficulties associated with the sharing and diffusion of synthetic and symbolic knowledge. The way the difficulties are dealt with has been widely studied under the guise of a social and practice based epistemology of learning (see for example, Amin and Cohendet 2004; Brown and Duguid 1991; Orlikowski 2002). In this approach organizational learning and interactions between indi-

Table 2.5 Knowledge-bases and their different characteristics and modes of spatial diffusion

Knowledge base	*Analytical*	*Synthetic*	*Symbolic*
Key characteristics	Solutions found in scientific models or equations	Solutions developed by applying or combining existing knowledge	Solutions based on hard to explain tacit insights
Exemplary industry	Biotechnology	Furniture manufacturing	Film directors
Means of sharing and diffusing knowledge	Publications The internet	Face-to-face interaction	Hard to share or diffuse Developed in practice over time and possessed by key individuals

Source: Asheim *et al.* 2007.

viduals in different offices of a global agency's network are associated with collaborative and constructive engagements which involve 'open experience sharing, where members articulate and exchange perspectives and intellectual resources and take others' opinions into consideration … and open dialogue, through which members shape meanings and develop "shared repertoire" … by reconciling differences and negotiating working consensus' (Hong *et al.* 2006: 412). Drawing on the approach developed in work on communities of practice (Lave and Wenger 1991; Wenger 1998), the social and practice based epistemology sees transnational networks such as those developed by global agencies as helping form a community of workers that together learn about ways of producing advertising for a global client. This learning occurs through interactions and participation in shared spaces of work. As Jones (2009) describes, these spaces of work are constituted through interactions occurring in material spaces that involve co-presence – face-to-face project meetings in one office – but also virtual spaces constructed through interactions involving the mediums of email, telephone and videoconference (see also, Faulconbridge 2006; Faulconbridge *et al.* 2009). In addition, as Beaverstock (1996, 2004) has shown, on some occasions the transnational forms rely on expatriates who, because of their symbolic knowledge, need to spend extended periods working away from their home office in order to apply their knowledge to a particular project and support the activities of other offices in the firm's network.

The most significant thing about the transnational agency form relying on social and practice based learning is the different approach taken to advertising compared with the 'imperial' export model. The agency's global network is a device for managing multi-directional relationships that involve several offices working together to develop advertising for a global client rather than one office acting as the command and control point for a campaign. As such, reflecting the arguments of Jones (2002), the power relations between offices are reconstructed by the transnational organizational form as a result of the need for multiple offices to collaborate rather than simply comply with orders issued by one office. Debates about power relations in TNCs are complex (see for example, Allen 2003; Bartlett and Ghoshal 1998; Dicken *et al.* 2001; Hardy 1996) and we return to the issue of power again in the next chapter. This initial discussion of power does lead us, however, to the second role of the network embeddedness of global agencies, brand management.

One of the principal roles of the transnational organizational form, alongside its role in innovation, is the coordination of worldwide campaigns so as to enable the brand 'stewardship' or 'guardianship'. Collaboration is important in the transnational agency and plays a central role in allowing brand 'stewardship' or 'guardianship'. This is particularly because social practice based processes of learning and interaction between workers in different offices allow, as much as possible considering the spatial entanglements affecting consumer–product relationships, consistency in the way a client's brand or product is promoted. The nature of such integrated transnational brand 'stewardship' or 'guardianship' is the subject of Chapter 5.

Conclusions

In sum, what is most significant about the transnational organizational form of the global advertising agency in the twenty-first century is the difference that exists in the geographies of the advertising production process when compared with the agency of the mid twentieth century. The export model of the mid twentieth century that located production of adverts in a few offices has been replaced by a collaborative process that involves multiple offices in the production process, thus giving more and more offices and cities a strategically important role in an agency's work. In this chapter we have documented the broad transitions within the advertising industry that drove this change, noting in particular how the emergence of a reflexive, spatially heterogeneous consumer market necessitated advertising that responded to the spatial entanglements which affect the relationship between consumers and a brand or product. To deliver such advertising requires an advertising production process that is societally embedded, something made possible by using office networks as tools for understanding situated consumer behaviours.

In the next chapter we consider the implications of such developments for the role of cities in advertising globalization. Specifically, we further consider the territorial embeddedness of agencies and the way this relates to the forms of societal embeddedness outlined in this chapter and then explore how such territorial embeddedness influences the geography of advertising work and the power relations that underlie transnational collaboration in the advertising production process in global agencies.

3 Cities and the grounding of global advertising work

Perhaps one of the most important debates in relation to knowledge-intensive business services and globalization has centred on the role of cities in the post-industrial knowledge economy. The topic has been the subject of books (see for example, Bryson *et al.* 2004; Daniels 1993), most notably under the guise of work on world or global cities (see for example, Hall 1966; Sassen 2000, 2006a; Taylor 2004; Taylor *et al.* 2006). Here our interest lies in analysing the way global advertising agencies 'use' cities as part of their contemporary globalization strategies. To do this analysis, we draw on work on knowledge-intensive business services and world cities, and also clusters, regions and, because advertising can be considered to be a creative industry, the relationship between creative cultural economies and cities. The latter especially has been the subject of numerous books (see for example, Cooke and Lazzeretti 2008a; Power and Scott 2004; Scott 2000), with some becoming key policy handbooks for regional economic development (Florida 2002; Landry 2000; Porter 1998). In summarizing this extensive literature, the aim is to reveal the processes by which cities become strategic basing points for global advertising work, and use this to explain the geography of advertising work in, and through, cities in the twenty-first century. Adopting the terminology introduced in the previous chapter, we will consider how global advertising work is territorially embedded and the implications of this for different cities, and their strategic role and powerfulness in global agencies' networks. Indeed, whilst we do not repeatedly use the phrase territorial embeddedness in this chapter, most of the arguments are explicitly about the way global agencies are territorial embedded in cities.

Regions, cities and economies

Work analysing the role of cities and regions in the global knowledge economy is an important starting point for any discussion of the role of cities in advertising globalization. The work of Michael Storper is particularly significant in this regard (Storper 1997; Storper and Salais 1997), providing one of the key interventions in debates about the de- and re-territorialization of economic activities in the contemporary period of globalization. Storper's (1997) arguments about the different 'worlds of production' have been rehearsed extensively (see for example, Bathelt and Glückler 2003; Cooke and Lazzeretti 2008b) and here we extract the points most relevant to our interest in advertising globalization. For Storper (1997), cities and regions have experienced a resurgence in their importance in the late twentieth and early twenty-first centuries because of the post-Fordist regime of accumulation. Specifically, Storper (1997) highlights how the logics of flexible specialization and consumer responsiveness which underlie post-Fordist regimes, and the associated time sensitive nature of production, generate modes of production reliant on resources such as labour, knowledge and suppliers that are most easily accessed through presence in agglomerations and localizations of business activity in particular cities or regions. It is, therefore, worth unpicking in more detail the nature of such agglomeration and localization processes so as to consider their usefulness in explaining the way particular cities 'ground' global advertising work. Underlying work on agglomeration and localization is one common argument: that advantages accrued from presence in particular cities or regions allow access to resources that facilitate innovation and the production of bespoke advertising products for clients.

Localization, innovation and creativity

As a concept, localization is usually used to refer to the knowledge related benefits acquired from co-location with firms operating in the same industry, although Glaeser *et al.* (1992) remind us that such benefits may also be accrued from co-locating with unrelated firms operating in different industries. Specifically co-location in a city or region is associated with the accrual of benefits both from forms of collective learning – processes through which firms learn from other colocated firms – and from access to pools of skilled labour which form in cities or regions. Debates about the nature and value of such localization economies are not particularly new. Marshall (1952) was first to note that specialized regional economies provide opportunities for industry-specific collective learning. He argued that in industrial districts, 'so great are the advantages which people following the same skilled trade get from near neighbourhood to one another ... [t]he mysteries of the trade become no mysteries; but are as it were in the air' (Marshall 1952: 225). More recently work has focused upon how the learning occurring in localization economies produces tacit knowledge which, 'because it is difficult to transfer ... may constitute a basis for sustained regional competitive advantage' (Lawson and Lorenz 1999: 306). Or as Tallman *et al.* (2004: 268) suggest, localization economies provide competitive advantage because,

> regional clusters indeed do possess certain competencies that provide competitive advantage to their constituent firms as a group ... Part of a regional cluster's advantage in its industry is tied to component knowledge that has originated within the cluster and remains there.

Locating in particular regions and cities has, therefore, been associated with processes of innovation and creativity. When firms with what Boschma (2005) calls 'related variety' – shared knowledge-bases and products – operate in proximity to one another they are said to gain access to knowledge that allows incremental innovations. For cultural industries such as advertising this is particularly important. Scott (2008) suggests that as part of a 'cognitive cultural economy', industries such as advertising, architecture, music and fashion are reliant on both intellectual, and also affective forms of knowledge which allow responsiveness to the needs of reflexive, individualized and spatially entangled consumers. Such expertise, according to Scott (2008), depends both on the skills of individual workers, and also the generative effects of urban milieu that help individuals develop the type of tacit, encultured and embrained knowledge associated with the production of effective adverts, and other cultural goods such as music, that have their value determined by subjective and hard to explain and predict consumer responses. Or, to put another way, cultural industries such as advertising deal with what Allen (2002) calls 'expressive symbolism'. Describing such a phenomenon, Allen (2002: 460) suggests:

> Expressive symbolism is perhaps best understood as a structure of feeling where, for example, a stylish piece of fashion or the lyrics of a new musical composition 'move' us in some way that is unrelated to, say, the latest 'language' of fashion or the technical competence by which the music is reproduced ... In common with all forms of aesthetic knowledge, an appreciation of film, art, design, music, display, and others rests upon their sensuous form the feelings they express not simply upon their technical or analytical excellence. In short, there is a creative content to such affects that cannot readily be measured by any abstract yardstick.

The expansive body of work that connects cultural industries to cities and their localization economies (see for example, Cooke and Lazzeretti 2008a; Power and Scott 2004; Pratt 2000; Scott 2000) identifies three mechanisms associated with region-specific forms of collective learning that facilitate the production of the tacit, encultured and embrained knowledge needed to succeed in cognitive cultural/expressive symbolic economies.

First, the term 'buzz' has been used to capture the role of the social interactions and gossip, rumour and discussion of industry-specific topics that occurs when employees from competing firms work and socialize in one city or region. For both Morgan (2004) and Storper and Venables (2004), such buzz relies on physical proximity between workers because of the way proximity facilitates frequent face-to-face encounters, the development of trusting and reciprocal relationships and the emergence of city or region-specific industrial languages that allow sense to be made of common business challenges. Empirical studies have shown that such face-to-face encounters and the development of trusting relationships based on shared industry languages occur in a range of forums as diverse as the professional association (Benner 2003; Faulconbridge 2007a), the trade show (Rantisi 2002) and the bar (Thrift 1994), and allow the exchange of opinions and ideas in a way that mirrors the social practice based mechanism of learning described in Chapter 2.

Second, localization has been shown to allow learning through non-verbal mechanisms such as observation. For example, Henry and Pinch (2000) show how GP Formula 1 manufacturers based in the Motor Sport Valley (in the English Midlands) cluster and benefit from seeing rivals' cars on test tracks and hearing the sound of engines and gear changes. Similarly, Rantisi (2002) reveals how fashion designers in New York City benefit from shopping in rivals' stores and seeing and touching their new designs. In both the case of Formula 1 car manufacturers and fashion designers, inspiration gained from such observation helps 'feed' the innovation process as rivals' products are not copied, but used as a springboard for the development of new designs.

Third, the presence of several firms with related variety in one city or region is also associated with the development of an expert labour pool that can be exploited by firms. Firms operating in localization economies have been shown to benefit from simplified recruitment processes because of the ease of movement of workers between firms in a city or region. The benefits of such intra-regional labour churn have been described vividly in terms of the *international* competitive advantage developed by regions and cities such as Motor Sport Valley (Henry and Pinch 2001), Silicon Valley (Saxenian 1994), New York City in relation to garment production (Rantisi 2002) and Los Angeles, New York City and Nashville in relation to music (Scott 1999).

In terms of our interest here, such localization processes have been shown to play a crucial role in supporting the development of the type of industry-specific knowledge that individuals need to be successful in the cognitive cultural economy of advertising. Advertising work, including the work of global advertising agencies, has been intrinsically associated with urban locations and during the latter half of the twentieth century became tied to a number of clusters within major cities such as London and New York. As Leslie (1997a) notes, in the case of New York City advertising agencies traditionally clustered in Midtown around the thoroughfare of Madison Avenue. However, by the mid 1990s, agencies had begun to 'abandon Madison Avenue' in favour of new locations in SoHo and Greenwich Village where rents were lower. Despite this move, as Leslie (1997a: 583) notes, firms still wanted to be in proximity to one another:

> The dispersal of agencies away from Madison Avenue does not suggest the diminishing significance of place to the operation of advertising. Rather, when agencies move they relocate in dense nodes of exchange. This suggests advertising agencies need to locate close to their competition.

New clusters began to emerge further south and west on Manhattan Island as firms sought to reproduce the buzz of Madison Avenue and, in particular, the interpretative value of collective learning about advertising challenges.

A similar story can be told in relation to London. As Grabher (2001: 352) describes, Soho in London acts as a 'massive spatial concentration of advertising agencies and, in fact, of the entire chain of activities associated with advertising, ranging from graphic design, lithography, photography and music to film and production and post-production in roughly one square mile'. This spatial concentration allows labour churn, both between agencies, but also between agencies and related firms such as those involved in design or photography, and at the same time has the same role as the clusters found in New York City in terms of collective learning. Indeed, as Pratt (2006) outlines, agencies in London use the buzz of the Soho cluster as a key tool for assessing their own and others' adverts and learning about the most successful campaign techniques. As Pratt (2006) demonstrates, 'peer regard' – the response of fellow advertisers to a campaign – is vital for understanding why a campaign is well received (and wins awards) or is derided within the industry (see also Faulconbridge 2007a). Consequently, in Soho the concentration of agencies forms what Grabher (2001: 371) calls a 'village' that:

> does not lead to a quasi-epidemic spread of a hegemonic 'one best way' but rather triggers (agency-)specific ways of adoption, recombination, or outright rejection ... spatial proximity in the Village fuels rivalry, that is, an ongoing engagement with different ways to organise, label, interpret, and evaluate the same or similar activities.

For Storper (1997), such combinations of collective learning through buzz and observation and the labour related benefits of regional and city based localization economies together forms a set of 'untraded interdependencies'. Firms benefit thanks to the access to knowledge that the co-presence of rivals and firms with related variety facilitates, something that is not reliant on trade related exchanges between firms. Johnson (2002) describes such benefits as a form of 'emergence', defined as an organic process involving the interaction of multiple actors that together and unintentionally generate creativity and innovation. Such 'emergence' is one of the ways that the work of knowledge-intensive business service firms such as advertising is territorially embedded in particular regions and cities. Indeed, a similar logic is implicit within two of the most influential models of city based work to emerge in the past twenty years, the cluster theory of Porter (1998) and the creative class theory of Florida (2002).

For both Porter and Florida successful cities and regions are built upon their resources. Porter emphasizes the role of clusters of firms – defined as 'geographic concentrations of interconnected companies, specialized suppliers, firms in related industries, and associated institutions (for example universities, standards agencies, and trade associations) in particular fields that compete but also cooperate' (Porter 1998: 197–8). Porter highlights clustering through his 'diamond model' of how the competitiveness of a cluster stems from vertical and horizontal inter-firm linkages and the information flows, market awareness and knowledge these produce. In addition peer pressure/competition from co-present firms is said to also drive innovation. As such Porter ascribes to the logic that co-location leads to innovation through collective learning.

For Florida (2002), the competitive advantage of a city is a result of subtly different processes, being reliant primarily on the abilities of the labour force attracted to a city or region. Florida pays less attention to the role of collective learning and emphasizes the value of individual talented workers, meaning his prescription for the successful city region involves developing the type of work-life environment that attracts talented workers. His 'three Ts' approach emphasizes the need for talent, in the shape of expert workers, tolerance, in the form of a diverse and tolerant community, and technology, the infrastructure and industries that allow entrepreneurialism, if a city or region is to be competitive. The presence of the 'three Ts', Florida (2002) argues, leads to a self-reinforcing process whereby talented workers are drawn to the city or region and, as a result so are firms.

There are many criticisms of both the work of Porter (1998) and Florida (2002), both in terms of the logic and politics of their models (see for example, Benneworth and Henry 2004; Markusen 2006; Martin and Sunley 2003; Peck 2005). We return to these critiques towards the end of the book in light of the findings of our empirical analysis. Here, we simply draw attention to the way the work of Michael Porter and Richard Florida complements existing understandings of the role of localization economies.

Agglomeration, projects and demand

Localization advantages are not the only assets that have been highlighted in studies of the way cities or regions territorially embed work in the global knowledge economy. Agglomeration advantages have been used to refer to the benefits accrued from the co-location of firms in different and possibly unrelated industries. At its simplest, agglomeration refers not to knowledge, creativity and innovation related benefits, but to savings made in relation to the cost of other resources (public transport, information communication technology) when firms co-locate and share key infrastructures. However, agglomeration can also bring other benefits for both manufacturing and knowledge-intensive business service firms. Two benefits, the co-location with clients and members of 'project ecologies', are especially important for business services such as advertising

The emergence of post-industrial urban service economies, the hallmark of the uneven city based geography of knowledge-intensive business service work in the late twentieth and early twenty-first centuries, has been extensively analysed (see for example, Allen 1992; Bryson *et al.* 2004; Daniels 1993; O'Farrell and Hitchins 1990; Wood 2006). In this work agglomeration advantages have been shown to be tied to the need for regular interaction between knowledge-intensive business service firms and their clients. This relates to previous arguments made about the way such firms provide bespoke advice in the form of solutions to clients' problems that are informed by the tacit expertise of the employees of the business service firm. Clients find it hard to assess such advice because they often lack knowledge of the particular issue, whether it be legal, accountancy or advertising related, they are being advised about. Hence, this is one reason why they turned to the knowledge-intensive business service firm for advice. Consequently, regular face-to-face meetings between clients and individuals representing the knowledge-intensive business service firm providing advice are vital (Jones 2007). These meetings allow, in the first instance, the client's needs to be fully understood by representatives of the firm and a trusting relationship to be developed between the client and the individuals providing the advice. As any project proceeds further face-to-face meetings that allow the delivery of advice to the client are vital both to ensure clarity of communication and, again, to reinforce the trust based relationship between the client and individuals representing the knowledge-intensive business service firm.

As a result, agglomeration and co-location with unrelated industries is beneficial for knowledge-intensive business service firms when this means a large number of potential clients are present in the same city or region. The time saved on travel when clients are located in the same city or region and the ability to arrange meetings at short notice provide competitive advantages lost when located outside an agglomeration economy. In turn the meetings facilitated by co-presence with clients then help inform and speed up the innovation process because of the insights gained into the client's business challenge, and the benefits accrued from accessing the client's understanding of the suitability of proposed solutions. Such insights are all accessed more easily when trusting relationships are constructed quickly and effectively between the client and the firm providing advice.

For advertisers, such agglomeration advantages are crucial during all stages of the production process. Pratt (2006: 1886) describes the five core stages of the agency–client interaction process as follows:

1 The 'pitch', when multiple agencies present their ideas and vie to win the right to work for the client.
2 The 'sign off' stage, which involves multiple meetings as the client reviews and agrees to their chosen agency's campaign proposal.
3 'Working up' the idea, a process that requires client responses to detailed strategies and mock-ups of the creative work to be used on a campaign.
4 Decision on content and campaign, when the final strategy is signed off by the client.
5 Making the advert, a process that involves the client less but requires, at minimum, a meeting to approve the final advert.

Grabher (2004) also recognizes how such agency–client relationships are important, in particular because agencies have to earn the respect of clients, being service providers like any other knowledge-intensive business service firm, not autonomous artists. All in all this means being in proximity to clients is a major advantage advertising agencies accrue from presence in an urban agglomeration economy.

In addition, agglomeration economies have also been shown to be advantageous because of the growing importance of project based working in a range of industries. The now well documented proliferation of the use of temporary project teams in knowledge-intensive business services as diverse as architecture and law, engineering and music production, has been highlighted as another feature of post-Fordist production regimes (see for example, Scott 1999; Whittington *et al.* 1999). In particular, the importance of innovation and the exploitation of the expertise of skilled individuals as part of attempts to deliver bespoke and novel products to clients and gain competitive advantage in the marketplace has been shown to be behind the trend towards a mode of organization of work that involves a constant churning of employees between temporary project teams established to complete one-off, time-limited projects. Such project teams have also been shown to blur or render fuzzy the boundaries of the firm. Whilst many project teams comprise individuals from different departments of one firm, more and more project teams, 'constitute an organizational form of coordinating activities and relations among legally autonomous, but functionally interdependent firms and individuals' (Sydow and Staber 2002: 216). In such scenarios project teams comprise members of a lead firm, but also freelance individuals hired to work on the project because of their unique skills, and/or representatives of a separate firm whose services are purchased for the duration of the project. For example, Scott (1999: 1968) provides a detailed breakdown of the project team involved in producing a new record in the music industry, listing twelve different groups of individuals, most of whom are not employed 'in house' by the recording company leading the project but operate as freelancers or independent companies (Table 3.1). Storper and Christopherson (1987) offer a similar list for the motion picture industry in Los Angeles and, as Table 3.1 also outlines, advertising is reliant on a similarly long list of project team members who are drawn from a range of related industries.

There is now an extensive literature on projects and such temporary teams. This literature covers issues including the way project teams and the decisions made in them are influenced by organizational histories and contexts (Engwall

Table 3.1 Members of a project team employed by a recording company producing a new music record and advertising agency developing an advert

Music	*Advertising*
Artists	Actors
Song writers	Talent agencies
Talent agencies and artist managers	Graphic design
Producers	Lithography
Sound engineers	Photography
Recording studios	Film production
Musical instrument suppliers	Musicians
Musicians	Set designers
CD manufacturers	Post-production editing (photography, music and film)
Promoters and distributors	Media services
Lawyers	Market research
Music publishers	Lawyers

Source: for the music industry, Scott (1999: 1968); for the advertising industry Grabher (2001) and the authors' research.

2003), the role of institutional thickness and shared social conventions in the facilitation of project based work (Sydow and Staber 2002) and the dangers of projects in terms of their tendency to inhibit organizational learning and their failure to capture 'lessons learned' (Hobday 2000; Grabher 2004; Grabher and Ibert 2006). The latter is because of the constant churning of individuals between teams and the tendency of workers to develop their own personal networks and contacts instead of developing institutional, firm assets and networks that can be exploited in the future. Of most interest to us here is existing work that highlights the spatial dimensions of project teams and the role of cities in facilitating project work.

The increasing reliance on project teams made up of individuals drawn from within and outside the organization's boundaries has rendered cities and regions important strategic sites for project work for several reasons. Cities and regions act as agglomerations both of freelancers and firms that can be drawn on as and when needed to provide particular expertise that is crucial for a project's success. At one level this means the process of searching for an individual or firm with the required expertise is simplified. As Vinodrai (2006) shows, cities and regions provide an 'ecology' of labour made up of individual workers, labour market intermediaries and multiple co-located firms from the same industry. This ecology produces career paths for a cohort of freelance and temporary workers built on multiple short-term contacts, something problematic for many workers who lack the security of a long-term contract (Ekinsmyth 2002), but beneficial for firms seeking to 'buy in' knowledge and expertise as and when needed. This is, however, only part of the benefit of locating in an agglomeration economy when using temporary project teams. At another level, the advantage of presence in particular cities or regions relates to the institutional embeddedness of project work. As Sydow and Staber (2002) outline, projects comprising freelancers or individuals representing firms autonomous from the project's lead firm are reliant on:

1. Regular face-to-face meetings. Like the relationships between a knowledge-intensive business service firm and its clients, meeting face-to-face is vital both to develop trust between parties and to enable effective communication.
2. Repeat relationships and the development of an institutional structure for project work. This institutional structure develops over time through the face-to-face interaction associated with project work and leads to social norms (around trust, reciprocity, commitment etc.) being shared by individuals which help smooth the project management process.

As such, in the language of economics, presence in the agglomeration economies of cities or regions minimizes the transactions costs associated with project work.

Cities and regions are also important sites for project work because of the benefits that co-presence within an agglomeration of freelancers and firms providing project services brings for quality control. As Grabher (2002) argues, such quality control need not be a deliberative process when located in an agglomeration economy. The 'noise' firms are exposed to, in the shape of rumours, recommendations and impressions about the work of potential project members, allows an automatic quality control process to occur. Exposure to this noise allows assessments to be made about the suitability of an individual or firm for any particular project whilst also, at the same time, working in reverse and allowing freelancers and firms providing project related services to quickly learn about opportunities for new work. When coupled together with the labour related advantages described earlier, Grabher (2002) suggests cities provide the ideal 'ecology' in which project work can be effectively executed.

Returning to Table 3.1, the vast array of related industries that support the development of adverts means that advertising agencies benefit greatly from presence in cities such as London and New York. Indeed, one of the responses of global agencies to the challenge of the boutique has been a refocusing around the core competencies of campaign planning and design with other tasks (for example, artwork or film editing) being outsourced to allow agency staff to focus exclusively on the strategic and creative planning of the campaign. For advertising, such project work is, therefore, central to the organization of the contemporary global agency. Grabher (2004) offers a detailed analysis of such advertising project working, showing how:

1 Personal networks are vital in the advertising industry, allowing individuals to be targeted and recruited by those running a project team.
2 Projects allow originality and rupture (i.e. creativity) because of the way they allow different individuals with different perspectives and skills to be brought together on every project.

As such, cities and regions are important in advertising project work because of the benefits of 'being there' in terms of the management of the extra-firm dimensions of knowledge-intensive, innovative production. 'Being there' ensures the right project team members are chosen and the relationships between the project team members are as effective as possible in terms of collaboration, trust and ultimately innovation.

There is also another dimension to the role of cities and regions in project work that has been somewhat underplayed in existing literatures. As Grabher *et al.* (2009) argue, the role of the consumer in innovation processes has been generally underestimated (although see von Hippel 1978, 2005). Studies of user-led innovation have begun to rectify this problem (see for example, Jeppesen and Molin 2003) and here, following Grabher *et al.* (2009), we suggest that the geographical insights of such work deserve further attention in the context of discussions of the role of cities and regions in projects. Specifically, we argue that for certain industries the consumer is as important as the client in the innovation process. So for advertisers, but also designers (Sunley *et al.* 2008) and computer game manufacturers (Johns 2006), it is important that the client commissioning a piece of work, but also, and perhaps more importantly, the consumer has their needs met by the final product. Meeting the needs of the consumer means developing an in-depth understanding of their consumption patterns, everyday behaviours and identities. The user-led innovation literature suggests such understanding can be developed using a range of strategies, ranging from direct interaction with and observation of consumers at its simplest (watching and learning from them), to co-development through consumer communities that contribute design ideas, test prototypes and offer feedback and actively play a role in the project team as designers.

Because certain cities and regions act as concentrations of consumers, locating in these places allows many of the types of interactions between firms and consumers that allow user-led innovation. So as Grabher *et al.* (2009) describe, presence in regions and cities is beneficial for user-led innovation because it makes consumer interaction and observation easy. It is straightforward to organize a focus group or simply watch consumers at work and play when in a bustling metropolis. It is also relatively simple to arrange more advanced forms of user-led innovation and co-production such as consumer communities which can be quickly formed and can meet regularly in cities.

For advertising as a cognitive cultural economy such consumer inputs have become increasingly important. The reflexivity of consumers and their spatial entanglements requires advertisers to develop ever more sophisticated strategies and engage in a form of user-led innovation in which the consumer is integrated into the advertising production process rather than treated as an unresponsive target object. As such, considering the consumer as part of the innovation process and project ecology provides another explanation of why cities and regions are important in the global knowledge economy and particularly important for cultural industries such as advertising. As Daniels (1995) notes, London, New York and Tokyo have developed as a 'golden triangle' of advertising work precisely because of such localization and agglomeration assets.

However, it is important not to over-emphasize the role of localization and agglomeration economies. Reflecting a wider critique of work on localization and agglomeration economies, Grabher *et al.* (2009: 256) argue that user-led innovation and co-development, 'is rarely confined to the more familiar dynamics of agglomerated systems of innovation. Rather, we expect co-development to enhance our understanding of the forms and functions of circulation of experts, knowledge, and artefacts'. Similar claims have also been made about the global reach of processes of collective learning (Bathelt *et al.* 2004; Grabher 2001; Faulconbridge 2006) and project ecologies (Grabher 2004), although Nachum and Keeble (2000: 20) do contend that, '[i]nternal TNC linkages are no substitute for locally generated knowledge needed for production coordination and integration. The need to acquire this type of knowledge encourages location of foreign and indigenous firms alike'.

It is to the 'global' spaces of advertising work and innovation that we now turn our attention. This is important both because it would be misleading to suggest that the assets of cities and regions described so far are exclusively territorially embedded – i.e. only accessed via physical presence in a particular place – and because our interest in the role of cities in advertising globalization and the work of global agencies inevitably means considering how cities generate and ground global flows of advertising work, labour, knowledge and capital.

Globalization, cities and spaces of flow

Perhaps one of the most fundamental developments in the theorizing of cities in the last twenty-five years has been the development of a conceptual framework focusing on world and global cities research (we use the former terminology from hereon in), which deals with the role of cities in, and reproducing, globalization. Since the world city hypothesis of Friedmann (1986) an expansive body of literature has explored the way that, contrary to the expectations of many, globalization has 'renewed the importance of major cities as sites for certain types of activities and functions' (Sassen 2000: 4), producing intense concentrations of economic activities in world cities at a time when information communications, were said by some (O'Brien 1992) to be leading to the 'end of geography' and a declining role of places generally, and cities specifically, in the economy. Numerous writings provide a comprehensive exposition of the theoretical explanations of world/global cities (see for example, Brenner and Keil 2005; Sassen 2000, 2006a; Taylor 2004; Taylor *et al.* 2006) and here we extract the most significant points that are relevant to our discussion of the grounding of advertising work and processes of innovation.

Table 3.2 indicates a range of factors important to understanding the assets that cities provide for knowledge-intensive business services (also known as advanced producer services) such as advertising. At its most basic, the advantage of cities to firms is that they constitute two sets of assets, internal and external. This enables city economies to expand in two ways: city agglomeration economies and city network externalities. The former are encompassed in the localization and agglomeration advantages discussed previously, whilst recent world city research, summarized in Table 3.2, has highlighted the relative network advantages of cities. This work moves away from the original 'world city hierarchy' conceptualization of Friedmann (1986) and, drawing on Castells' (2000) concept of the network society, emphasizes instead world city networks (Beaverstock *et al.*, 2000; Taylor 2004) as the defining feature of strategically important cities in globalization.

The world city network has been specified as an interlocking network model, where knowledge-intensive business service firms (such as advertising agencies) 'interlock' cities through their multi-city office networks (Taylor 2001). Through the agency of their routine professional, creative and financial work, it is assumed that these firms generate knowledge flows (embodied and virtual) between their city offices, thus linking cities together through their myriad business projects. This is the network embeddedness referred to in Chapter 2. By collecting data on the size and functions of offices across cities, work relations in

Table 3.2 The production of the assets of cities, as understood by work on globalization and world cities

City asset	*Concepts and literatures examining the role of global flows*	*Conceptual argument*
Knowledge that informs innovation	Global buzz and pipelines (Bathelt *et al.* 2004). Global communities of practice (Amin and Cohendet 2004; Faulconbridge 2006, 2007b). Trade fairs (Maskell *et al.* 2006).	Cities tap into knowledge flows produced both by virtual interactions between individuals located in different cities and travel which allows occasional face-to-face encounters, for example at meetings or trade fairs.
Labour pools	Business travellers and expatriates (Beaverstock 2004; Faulconbridge *et al.* 2009). Transnational communities and class (Bunnell and Coe 2001; Sklair 2001).	Skilled workers are mobile and cities can tap into their expertise by acting as temporary sites of work for such nomadic individuals.
Clients	The globalization and off-shoring of service work (Bryson 2007; McNeill 2008). Service exports and trade (Dicken 2007; UNCTAD 2004).	Clients engage in worldwide searches for providers of the most appropriate services (as defined by level of innovation, cost etc.), thus creating international flows of trade and investment in services).
Projects	Global project teams and collaborations (Grabher 2001; Orlikowski 2002).	Firms rely on project teams involving members in multiple offices worldwide with collaboration occurring through virtual (email, videoconference) and embodied (travel) flows.
User–led innovation	Online user communities and user-led production (Grabher *et al.* 2009; Mateos-Garcia and Steinmueller 2008).	User–producer co-production occurs virtually allowing firms in any one city to tap into flows of consumer expertise from throughout the world.

the world city network can be estimated to show the structure of the network. The key measure is network connectivity. This indicates the degree of integration of a city within the network. For instance, in nearly all world city studies, London and New York emerge as the two most integrated cities in this global servicing of business. Thus, for instance, for many financial service firms, there are massive network advantages to being in London and New York because the success of their business depends upon the myriad intersections of knowledge flows between these two leading international financial centres, compared, say, with flows between Frankfurt and Charlotte.

Research on the world city network has generally focused upon measuring 'network connectivity' through a focus on a range of service sectors: office networks in accountancy, advertising, financial services, legal services and management consultancy. Latest results of different cities' network connectivity (for 2008), based upon 175 office networks across 525 cities, show the top twenty network scores indicating the cities most intensely integrated into the world city network, and therefore, with the largest network (knowledge) assets (Table 3.3). The rise and importance of Pacific Asian cities is the main finding of this work (Derudder *et al.* 2010; Taylor *et al.* in press) but note the worldwide geography of these leading cities (except in Africa). Such results can be produced separately for the different sectors and results for advertising firms are discussed in the next chapter.

Table 3.3 The global network connectivity of cities, 2008

Rank	*City*	*Global network connectivity*
1	London	1.00
2	New York	1.00
3	Hong Kong	0.83
4	Paris	0.78
5	Singapore	0.75
6	Tokyo	0.74
7	Sydney	0.71
8	Milan	0.69
9	Shanghai	0.69
10	Beijing	0.69
11	Madrid	0.65
12	Moscow	0.64
13	Seoul	0.63
14	Toronto	0.63
15	Brussels	0.63
16	Kuala Lumpur	0.60
17	Mumbai	0.60
18	Buenos Aires	0.60
19	Chicago	0.57
20	Warsaw	0.56

Source: Taylor *et al.* 2010b.

The meaning of such research for understanding the importance of cities to advertising is that the role of cities such as London and New York is defined not only by what goes on in the 'Village', but also by what goes on in the networks of global agencies and the way this creates flows of knowledge, people, clients and user interaction into, through and out of the city. As Amin and Thrift (2002) describe, such flow and interconnectivity require a 'reimagining of the urban' so as to conceptualize cities not as bounded and discrete entities, but as connected, fluid and porous formations. Discussions in the previous chapter begin to shed some light on the nature of the economic flows that make up such networked cities and which reflect the transnational logic. For example, Faulconbridge (2006: 529–30) suggests global agencies connect territorially embedded communities to global communities through:

> a 'global practice based space' of learning. When the 'social production of new knowledge' is taking place focus falls on exchanging insights into globally common practices such as 'how to inspire feelings of desire towards a product' or 'how to revitalise the image of a product that is viewed by consumers as old fashioned'.

For Allen (2000) such conceptualizations of global spaces of learning connecting world cities together are vital as they recognize the way that learning associated with expressive symbolic or cognitive cultural knowledges occurs through topological practices that transcend scalar boundaries of the local or the urban. A similar story can be told in relation to the topological geographies of labour and project work. Global agencies operate as global labour pools themselves with expatriates moving between offices at intervals when their expertise is needed for a particular project (see for example, Millar and Salt 2008). Hence as Grabher (2004) suggests, projects are best seen as topological communities that can be reliant on stretched networks of collaboration involving individuals drawn from outside the city, as well as within the city's agglomeration economy.

Clients, as documented earlier, also generate global flows in agency networks. The original rationale for opening overseas offices, to serve global clients wherever they sold consumer goods, generated flows of capital into the overseas offices of global agencies which were used to employ local labour in different cities. With the development of the transnational organizational form, these flows have accelerated and, as a result, cities have benefited from more and more work being generated by network flows. User-led innovation, as Grabher *et al.* (2009) describe, need also not simply be a territorialized process. Whilst the spatial entanglements of consumers inevitably mean it is most fruitful to bring consumers from the 'local' market into the production process, for example consumers in London for adverts to be used in the UK, as Mattelart (1991) describes, much can also be learned from engagement with consumers 'at a distance' when degrees of commonality exist. Quoting a document from the agency Saatchi & Saatchi, Mattelart (1991: 52–3) highlights how consumers in the same city, such as those living in Midtown Manhattan and the Bronx, often have less in common than consumers in different cities and countries, for example those in Midtown Manhattan and the seventh Arrondissement in Paris. Of course, the different situated spatial entanglements of consumers mean that it would be too simple to suggest the same advertising campaign would work in both Midtown Manhattan and the seventh Arrondissement in Paris. But, this analogy does highlight how spatial proximity is not always the best means of engaging with consumers. Exemplifying this argument, 'web 2.0 generation' consumers are more and more familiar with internet applications that allow interactive content, virtual communities and online collaboration. Hence advertisers located in New York City, potentially designing campaigns that will run across the USA, need not limit themselves to user-led innovation involving consumers located in the same city but can increasingly rely on internet based spaces of flow as a tool for better engaging with, and responding to, consumer reflexivities.

All of the above lead to what Jones (2008) refers to as 'global work'. This is defined as work bound to a distanciated set of social relations facilitated by mobility and virtuality, and which generate spatial power relations between workers in different places and spaces. The concept of global work will be important in the rest of the book. Specifically we use global work to refer to work that agencies are involved in as a result of their spatial relationships with other offices in a firm's worldwide network, relationships built upon place-specific, but also network assets.

Territorialized and networked advertising cities

So what are the implications of the work on world cities for our understanding of the global geographies of advertising work? In the world cities literature, our reading of cities' strategic importance is a product both of what goes on in cities – the agglomerations and localizations – and also what flows through global networks and gets temporarily pinned down in a city. As such, it is the global practices of business elites in any city – their membership and participation in worldwide knowledge communities, their interaction with consumers via the internet – as well as their local practices that generate competitive advantage (see for example, Beaverstock 2007a, 2007b; Jones 2008, 2009).

So by going back to the basics of Jacobs' (1969, 1984) urban theorizing, world city research has firmly established that internal and external assets are complementary and together produce cities. Thus, the focus on networks in the world cities approaches cannot explain all of the processes involved in producing a successful world city, but when coupled to an understanding of what happens within cities is incredibly powerful. It allows, for example, answers to be developed to questions such as: how does a city become a key site at which mobile labour chooses to temporarily reside? Why do clients, in Frankfurt for example, choose to purchase goods and services from firms located in London? Answering these questions requires consideration of the way the territorialized resources described in localization and agglomeration approaches get both exploited and produced through global networks. Think, for example, of London or New York City. The two cities have long sat atop of the many city league tables produced both by academics and consultants (see for example Beaverstock *et al.* 2000; Faulconbridge 2004; Reed 1981; Corporation of London 2009) and this can only be explained with reference to both territorial assets – such as critical masses of co-located firms that form innovative knowledge-intensive business service localization economies with related project ecologies – *and* the way these assets are produced and exploited through global networks. A city strong in only network connections *or* territorial resources is likely to be far less powerful than a city able to effectively develop both.

For advertising, as an example of a cognitive cultural economic industry, the reliance on territorialized assets has received by far the most attention in existing literatures. In many ways this is unsurprising. The discussion in Chapter 2 on the reflexivity and spatial entanglement of consumers draws attention to the importance of situated advertising and the societal and related territorial embeddedness of the advertising production process. To produce successful adverts in the twenty-first century requires a spatially variegated approach that apparently lends itself to the exploitation of localization and agglomeration economies. However, as Grabher (2001, 2004) in particular has shown, such trends in the advertising industry should not draw our attention away from the simultaneous role for global networks which, when coupled to localized and urbanized production processes, create a topological map of advertising work, defined both by territorial assets and also network flows and connectivities. Such advertising communities require us to develop, in the words of Amin (2002: 386), 'a different interpretation, one which emphasises a topology marked by overlapping near–far relations and organizational connections that are not reducible to scalar spaces'.

So in terms of the question of how this understanding of both the city and network geographies of advertising production helps to explain the global geography of advertising, our analysis indicates that successful, powerful advertising cities manage to pin down a significant quantity of advertising work in the twenty-first century because of both their territorial assets that facilitate the production of effective adverts and, at the same time, their network assets that help reproduce and allow the most effective exploitation of territorial assets through global work. This is the archetypal story of the success of cities such as London and New York, but the conceptual framing developed here suggests it is equally the story of other cities that develop to differing degrees territorial and network assets that feed advertising work. Some empirical evidence that such a framing has value already exists. For example, Speece *et al.* (2003) point to the way Vietnam's advertising industry has changed over the past decade from being the home of global agencies' 'post box offices' to being the home of strategically important offices that produce adverts for the domestic market as part of the transnational agency model. Similarly Po (2006) highlights how the offices of global agencies in Beijing, Guangzhou and Shanghai are not simply reproducing US adverts overseas, but, instead, are developing local models of advertising based on the expertise of advertising workers in the three cities (see also Muller

2005). In both cases it is the ability of different cities to simultaneously develop localization and agglomeration assets – skilled labour forces and project ecologies that allow, in particular, situated adverts that respond to reflexive and spatially entangled consumers to be produced – whilst exploiting network resources – insights from individuals elsewhere in global agencies' networks, insights and the personal connections of expatriates or returnee workers who trained in the USA or UK, flows of demand from global clients selling consumer goods in the cities in question – that have allowed these cities to become new important sites of advertising work.

Conclusions

This chapter has drawn together the conceptual framing to explain the grounding of the globalization of advertising and global agencies in cities, and the role of the city's assets and network relations in reproducing the geographies of the global advertising firm. By highlighting the importance of understanding the contemporary geographies of advertising work as a process involving topological modes of organization that privilege both the city as a territorial formation and the network as a relational device, the discussion sets the scene for our empirical studies of advertising worldwide and in the three US cities of New York, Los Angeles and Detroit. As such, the rest of the book seeks to explore the implications of the understanding developed in this and previous chapters for explanations of the uneven geography of advertising work. This inevitably means exploring the successes of recent years – cities that have sustained or grown their role in the spatial divisions of advertising labour – and the failures – those cities that are playing a declining role in or have failed to pin down a significant proportion of advertising work.

In order to develop such an analysis, we begin Part II by examining the contemporary global map of advertising work. Specifically we seek to understand where advertising work 'gets done' in the twenty-first century, why it is done where it is and how understandings of the changing spatial power relationships associated with transnational advertising work can help in explanations. We then explore the work practices of the employees of global agencies and how these produce a global map of advertising work. In Part III we move on to examine our three case study cities in detail – New York, Los Angeles and Detroit – because of their exemplary role as cities that have sustained, grown and seen a decline in their importance in spatial divisions of advertising labour and work, and practice.

Part II

Geographies of advertising work in the twenty-first century

4 Cities and advertising globalization

New York, Los Angeles and Detroit in a global perspective

This chapter begins Part II by providing a macro-scale analysis of the geographies of advertising globalization. Through quantitative data we examine the geography of the work of global advertising agencies, the position of our three case study US cities in this geography, New York, Los Angeles and Detroit, and the questions raised about the forces (re)producing such spatial arrangements. We remind the reader at this point, that we use the term 'global agency' to describe agencies with offices in all major world regions and which use transnational campaign and organizational structures to deliver collaboration and cooperation between offices to develop advertising for a product or brand in multiple markets. But, also, at the same time, these global agencies form part of the connections and flows that help produce powerful world city networks.

American exceptionalism

The role of the advertising industry in contemporary globalization is a very special one. As already noted, advertising is an archetypal American service industry that grew in importance in the first half of the twentieth century as a new 'consumer society' was emerging. Advertising was an integral component of the 'consumer modernity' that American society was creating: it nurtured a growing band of mass consumers from needs to wants. Thus the advertising industry spread beyond the USA to become global in the final decades of the last century. But, the advertising industry was always more than just another case of globalization; advertising has been integral to global consumer society just as it had been to initial American consumerism. The advertising industry's job has been to channel modern expectations of a better life into concrete acts of choosing to buy, first American, and latterly global, brands. This is epitomized by Shanghai's famous, and wonderfully named, 'Mall of Global Brands', and this very particular modern history and role for the advertising industry has produced a quite distinctive global geography.

This geography has a very simple basic structure of two parts: the American heartland of the industry and the rest of the world. With its relatively long tradition of advertising work at the centre of the American economy, the USA has a mature and sizeable advertising industry which was, according to the US Bureau of Labor Statistics, at its peak in 2008 directly employing 185,000 people in agencies across the country with a further 280,000 employed in related services such as Public Relations. The leadership position of the USA in advertising was represented for many decades by 'Madison Avenue', the industry's 'home' in New York, and despite the dissipation of this cluster, more on which is in Chapter 6, the city remains the pre-eminent centre for advertising in the world. In other US cities advertising remains a vital service sector after nearly a century of building consumer society. In the rest of the world advertising has tended to become much more concentrated in just one city per country, it locates where the national TV, its main medium, is located. The result is a much denser geography of advertising in the sector's heartland, the USA, than anywhere else in the world.

This US exceptionalism is important when considering our three case study US cities within contemporary globalization and is the context for this chapter, which has two main purposes. First, the worldwide geography of advertising is described to provide the backstory of our qualitative empirical study of the globalization of advertising. Second, we locate New York, Los Angeles and Detroit within this context in order to understand our case studies in interrelated global and American terms. In this chapter we focus our attention on the first decade of the twenty-first century: benchmarking at 2000/01, 2004 and 2007/08 for detailed interrogation of advertising that leads directly into the case studies. We use three datasets to pinpoint what is happening. These are each very distinctive and are intended to provide an all round description of the geographies of globalization in the advertising industry.

1 Globalization and World Cities (GaWC) surveys of office networks to provide estimates of how cities are connected in advertising projects. These data are available for 2000 and 2008.
2 *Advertising Age* data on firms and markets to identify work, control and sales through cities across the world. Unfortunately this data is not available after 2001, somewhat limiting our analysis. Nevertheless, the data available does help paint an insightful picture of where advertising work 'gets done' in the contemporary period of globalization.
3 US Employment Census and Bureau of Labor Statistics data to provide information on jobs and agencies specifically for our three case study cities. These data are available annually and we use 2004 results to provide a snapshot of the mid-point in our period of analysis. We refer to more up to date data for our three case study cities in Chapters 6, 7 and 8.

Datasets One and Two are employed to provide a cross-sectional context for global advertising at the beginning of the twenty-first century. Dataset Three is used to focus in on our three case study cities and provide a comparative context that subsequent chapters can draw upon. The rest of the discussion in this chapter is, therefore, focused around fourteen tables that present our quantitative data with each our three case study cities capitalized, and other US cities are emboldened, to highlight cities that are discussed in more detail in subsequent chapters.

Advertising at the turn of the century

By 2000 advertising was an established global service industry straddling all continents albeit to different degrees. This geography can be envisaged in two related ways: as work involving inter-city networks, or as locales in which work takes place. We begin with the former.

Advertising agencies in world city networks – 2000

The Globalization and World Cities (GaWC) Research Network – more details of which can be found at www.lboro.ac.uk/gawc (accessed 1 September 2010) – carried out its first global survey of advanced producer service office networks in 2000. One hundred leading firms were studies and their worldwide office location strategies investigated. For each firm, the importance of their offices in 315 different cities was coded (from zero for no presence to five for headquarter office). More details of this methodology can be found in Taylor *et al.* (2002). The basic data consists of a 315 (cities) × 100 (firms) matrix in which each cell indicates the importance of a city within a firm's office network. Using this unique data matrix two sorts of results can be generated.

1 Network connectivity measures. These indicate the degree to which a city is integrated into the overall pattern of office networks. It is based upon the number of firms a city shares with other cities (potential project links) and the importance of those offices. These are reported as proportions of the highest score – the city most integrated into the networks – for ease of interpretation. In advertising New York always emerges as the most connected city and, therefore, other cities' connectivities are proportions of New York's connectivity. For details of computation see Taylor (2001).
2 Common location strategies. This data describes the locational strategy of service firms – where and how important their offices are across the world. These individual strategies can be compared and combined into different common location strategies. This is achieved using principal components analyses which search out firms that have similar location strategies (i.e. locate their offices in mostly the same cities). Component scores describe these common location strategies; they indicate how important a city is in a specific common location strategy. For further details see Taylor *et al.* (2004).

In the 2000 survey, seventeen of the 100 firms were advertising agencies and results based upon analysis of the locational strategies of these firms are highlighted here.

In Table 4.1 the top seventy cities are shown that all have network connectivities of at least 0.30 in 2000 (i.e. three-tenths of New York's connectivity). This long list is chosen so as to encompass all three of our case studies: Detroit with a score of 0.31 ranked only sixty-seventh globally. Two obvious things to note are the global dominance of New York that, as expected, was far ahead of all rivals,

Table 4.1 Global network connectivities (GNCs) of cities in the advertising industry, 2000

Rank	*City*	*GNC*	*Rank*	*City*	*GNC*
1	**NEW YORK**	1.00	36	Prague	0.39
2	London	0.79	37	Helsinki	0.39
3	Hong Kong	0.60	38	Zurich	0.38
4	Toronto	0.58	39	**San Francisco**	0.38
5	Sydney	0.57	40	Mexico City	0.37
6	Amsterdam	0.56	41	Shanghai	0.37
7	**Miami**	0.54	42	Seoul	0.37
8	Singapore	0.53	43	Jakarta	0.36
9	Milan	0.53	44	Budapest	0.36
10	Madrid	0.52	45	Stockholm	0.36
11	Frankfurt	0.51	46	Bangalore	0.36
12	Melbourne	0.50	47	Caracas	0.35
13	Tokyo	0.50	48	Bucharest	0.35
14	Paris	0.49	49	Chennai	0.34
15	Brussels	0.49	50	Guangzhou	0.34
16	Copenhagen	0.47	51	Karachi	0.34
17	Mumbai	0.47	52	**Atlanta**	0.34
18	Lisbon	0.46	53	Moscow	0.34
19	Sao Paulo	0.44	54	Bangkok	0.33
20	Taipei	0.44	55	Tel Aviv	0.33
21	Auckland	0.44	56	Cairo	0.32
22	**LOS ANGELES**	0.44	57	Sofia	0.32
23	Buenos Aires	0.44	58	Beirut	0.32
24	Barcelona	0.43	59	Dubai	0.32
25	**Chicago**	0.43	60	Panama City	0.32
26	Athens	0.43	61	Calcutta	0.32
27	Kuala Lumpur	0.42	62	Ho Chi Minh City	0.32
28	Johannesburg	0.41	63	Hamburg	0.32
29	Vienna	0.41	64	Guatemala	0.32
30	Warsaw	0.41	65	Santo Domingo	0.32
31	Beijing	0.40	66	Bogota	0.32
32	Manila	0.40	67	**DETROIT**	0.31
33	Montreal	0.40	68	Jeddah	0.31
34	New Delhi	0.39	69	Rome	0.31
35	Istanbul	0.39	70	Santiago	0.30

Source: authors' research and GaWC data (see www.lboro.ac.uk/gawc (accessed 1 September 2010)).

Note

Interpretation: figures indicate the degree of a city's integration into advertising office networks, as a proportion of the leading city's (New York) integration.

and, more surprisingly, the dearth of other US cities in the list: although Detroit has a low global ranking, among US cities it has the relatively high ranking of seventh. Los Angeles, ranking third in the USA in terms of billings, only appears in twenty-second position globally (the surprise of Miami ranking second in the USA on this global measure is discussed later). It seems that, as suggested above, the advertising industry has gone global but US cities (excepting New York) are not fully engaged in the process. This will be an important issue for discussion in Chapters 6, 7 and 8.

Let us be clear what this means. The firms included in the 2000 GaWC survey were all global agencies with a pattern of offices covering Pacific Asian, Western Europe and the USA. These are a small minority of all advertising agencies and their location strategies do not greatly encompass large numbers of US cities. This suggests that, for advertising agencies, 'going global' has meant producing an office network relatively evenly spread across the world to cover a myriad of 'national' markets. Thus Table 4.1 features forty capital cities plus other 'media primate' cities in countries where the capital city is not the leading media city (for example, Toronto, Sydney, Mumbai, Sao Paulo, Auckland, Istanbul, Zurich, Karachi and Dubai). Thus overall advertising agencies seem to mimic the international map of countries as they strive to cover different national markets, despite the problems with such 'scalar' strategies discussed earlier in Chapter 2.

We can obtain a glimpse of how this works in Table 4.2. This table derives from a principle components analysis of all 100 global service firms in the GaWC

Table 4.2 The imperial location strategy of advertising, 2000

Score	*Leading cities*	*Rank*	*Leading advertising agencies*
3.2	**NEW YORK**	1	Ogilvy
1.7	Johannesburg	3	D'Arcy
1.6	Helsinki	5	McCann-Erickson
1.5	Istanbul	6	FCB
1.4	Athens	8	Young & Rubicon
1.3	Toronto	10	Saatchi & Saatchi
1.3	Lisbon	11	J. Walter Thompson
1.3	Tel Aviv	12	BBDO
1.2	Madrid	13	Impiric
1.2	Copenhagen		
1.2	Brussels		
1.2	Buenos Aires		
1.1	Vienna		
1.1	Warsaw		
1.1	Sofia		
1.1	Manila		
1.0	Bucharest		
1.0	**Miami**		

Source: Taylor *et al.* (2004).

Note

Interpretation: the results are from a principal components analysis of 100 advanced producer service firms (including seventeen advertising agencies) across 123 cities worldwide. A varimax rotation was employed. Scores indicate how important a city is within a common office location strategy (only cities with scores over 1.0 are listed). The strategy is constituted by the office locations of the firms listed in the final column. Only advertising agencies that constitute the component are listed and their rankings among the 100 firms are given to indicate the dominance of advertising for this component.

2000 survey. The components define common location strategies and we highlight just one, which is the 'imperial' globalization strategy in which global firms seek to take control of markets in other cities (for further discussion see, Taylor *et al.* 2004). The scores in Table 4.2 indicate how important a city is in 'imperial' globalization strategies. Note the dominance of New York: this is deemed to be the articulator city for globalization strategies. Note also that seventeen different countries are represented by the other cities in the strategy: these cities (bar one, Miami) house offices serving national markets. The strategy has a worldwide distribution covering all continents but with Western Europe, beyond the four main economies (Germany, France, Italy, UK), particularly well represented. Finally, Miami, as a US city, appears to be an anomaly to this patterning, but this is not the case: in global servicing Miami often acts as the 'capital of Latin America' (Brown *et al.* 2004), which thus also accounts for the city's high ranking in Table 4.1.

The ranking of the agencies in Table 4.2 indicates their relative closeness to the derived common strategy: thus Ogilvy & Mather is the firm within the 2000 GaWC analysis that most closely matches this imperial strategy. The key point here is that nine of the top thirteen firms contributing to this strategy are advertising agencies; we can relabel it the 'imperial advertising strategy'.

What these GaWC analyses tell us, then, is that in 2000 the advertising industry was clearly global in scope, dominated by New York, and with clear evidence of an imperial structure articulated through New York i.e. agencies operating in, and their work in, New York were key drivers of the globalization of advertising. Neither Los Angeles nor Detroit featured prominently in this global analysis. We interpret this trend further in Chapters 6, 7 and 8.

City advertising markets

Until 2001 *Advertising Age* produced insightful rankings of cities by the size of their advertising market (as measured by billings generated in US dollars). This data is useful for benchmarking the role of different cities as locales of advertising globalization at the turn of the new millennium, but unfortunately cannot be followed up in Part III.

Table 4.3 shows the largest forty markets and, as before, New York is in first position by a long way. Note however that, compared to global connectivity, US cities feature much more prominently with Detroit and Los Angeles ranked sixth and seventh respectively. In all, fifteen US cities feature in this table. Although Table 4.1 and Table 4.3 are based upon different measures and, therefore, are not strictly comparable, the tables do suggest that when the restriction for advertising work being global (by office network) is relaxed, US cities come to dominate the world rankings. This is another example of the US exceptionalism referred to in the introduction, i.e. the fact that the US has multiple advertising centres. But it also shows how analyses of territorial assets alone – city based markets – can hide the role of network assets – connectivity into global space economies – in determining the nature and strength of a city's economy. Thus, not all US advertising cities are part of advertising globalization.

Table 4.3 Top forty city advertising markets, 2001

Rank	*City*	*Billings (US$ millions)*
1	**NEW YORK**	57,237.60
2	Tokyo	38,663.90
3	London	23,813.70
4	**Chicago**	15,212.70
5	Paris	12,440.90
6	**DETROIT**	8,336.80
7	**LOS ANGELES**	8,225.90
8	Milan	6,095.20
9	**San Francisco**	5,855.30
10	Frankfurt	5,512.70
11	Sao Paulo	5,480.10
12	Düsseldorf	4,743.60
13	**Boston**	4,156.10
14	Toronto	4,150.70
15	Amsterdam	4,121.30
16	Madrid	4,116.00
17	Sydney	3,959.60
18	**Minneapolis**	3,620.50
19	**Dallas**	2,457.00
20	Hong Kong	2,096.60
21	Istanbul	1,964.60
22	Mexico City	1,943.70
23	Brussels	1,882.50
24	Buenos Aires	1,670.20
25	**St. Louis**	1,651.00
26	**Atlanta**	1,619.60
27	**Miami**	1,473.40
28	**Austin, TX**	1,433.80
29	Athens	1,413.10
30	Hamburg	1,385.70
31	Vienna	1,379.90
32	**Kansas City**	1,329.50
33	**Philadelphia**	1,318.30
34	Mumbai	1,279.70
35	Copenhagen	1,275.20
36	Zurich	1,272.30
37	Taipei	1,202.70
38	**Columbus, OH**	1,201.20
39	Singapore	1,154.50
40	**Seattle**	1,044.10

Source: derived from data presented in Taylor (2008).

We can take this story of exceptionalism further by considering the top fifty advertising holding companies in Table 4.4. The companies are ranked and this list shows the great degree of concentration of income in 2001: the final column shows the ‘big three’ – WPP, Interpublic and Omnicom groups – and a top nine with a break in quantity of income after Hakuhodo. The continuing dominance

Table 4.4 Advertising holding companies ranked by gross world income, 2001

Rank	*Holding company*	*Headquarters*	*Worldwide gross income (US$ million)*
1	WPP Group	London	8,165.00
2	Interpublic Group of Cos.	**NEW YORK**	7,981.40
3	Omnicom Group	**NEW YORK**	7,404.20
4	Publicis Groupe (includes Bcom3 Group)	Paris	4,769.90
5	Dentsu	Tokyo	2,795.50
6	Havas Advertising	Levallois-Perret, France	2,733.10
7	Grey Global Group	**NEW YORK**	1,863.60
8	Cordiant Communications Group	London	1,174.50
9	Hakuhodo	Tokyo	874.3
10	Asatsu-DK	Tokyo	394.6
11	TMP Worldwide	**NEW YORK**	358.5
12	Carlson Marketing Group	**Minneapolis**	356.1
13	Incepta Group	London	248.4
14	DigitasA	**Boston**	235.5
15	Tokyu Agency	Tokyo	203.9
16	Daiko Advertising	Tokyo	203.0
17	Aspen Marketing Group	**LOS ANGELES**	189.2
18	Maxxcom	Toronto	177.1
19	Cheil Communications	Seoul	142.0
20	Doner	**Southfield, MI**	114.2
21	Ha-Lo Industries	**Niles, IL.**	105.0
22	Yomiko Advertising	Tokyo	102.2
23	SPAR Group	**Tarrytown, NY**	101.8
24	Cossette Communication Group	Quebec City	95.2
25	DVC Worldwide	**Morristown, NJ**	92.6
26	Clemenger Group	Melbourne	91.0
27	Rubin Postaer & Associates	**Santa Monica, CA**	90.3
28	Hawkeye Communications	**NEW YORK**	87.8
29	Panoramic Communications	**NEW YORK**	86.2
30	Richards Group	**Dallas**	84.5
31	Asahi Advertising	Tokyo	84.3
32	inChord Com. (Gerbig Snell/ Weishemer)	**Westerville, OH**	76.1
33	Bartle Bogle Hegarty	London	73.9
34	Wieden & Kennedy	**Portland, OR**	73.8
35	Cramer-Krasselt	**Chicago**	72.7
36	M&C Saatchi Worldwide	London	71.7
37	LG Ad	Seoul	67.6
38	Nikkeisha	Tokyo	66.5
39	AKQA	**San Francisco**	66.0
40	Armando Testa Group	Turin, Italy	62.9
41	Sogei	Tokyo	61.3
42	Springer & Jacoby	Hamburg	60.6
43	ChoicePoint Direct	**Peoria, IL**	59.4
44	Gage	**Minneapolis**	58.6
45	Harte-Hanks Direct & Interactive	**Langhorne, PA**	57.0
46	360 Youth	**Cranberry, NJ**	56.7
47	Ryan Partnership	**Westport, CN**	56.3
48	Envoy Communications Group	Toronto	54.8
49	MARC USA	**Pittsburgh**	53.2
50	Data Marketing	**Santa Clara, CA**	52.8

Source: derived from *Advertising Age* 2002.

of New York is shown through two of the 'big three' groups being headquartered there, plus another in the top nine, and three more in the top thirty. There is a Los Angeles company ranked at seventeenth, but no Detroit company listed. Note that in Table 4.4 globally, New York is joined by London, Tokyo and Paris (and Seoul to a lesser extent) in dominating headquarter locations and only five other non-US cities feature (Quebec, Melbourne, Turin, Hamburg and Toronto).

This US/rest of the world contrast is heightened further when the role of the top nine holding companies in Table 4.4 is considered in different cities worldwide. Table 4.5 shows the corporate concentration as a percentage of billings in a city through these companies in twenty-five key cities. The contrast is stark: beyond the USA global agencies totally dominate several key city markets whereas they play a smaller role in many US cities, including New York. Note, in particular, the case of Philadelphia. The score of zero does not mean that there is no advertising market in Philadelphia. But, it does mean the city's market is

Table 4.5 Corporate concentration by city

Rank	*Cities*	*% Billings by top nine companies*
1	Paris	99.6
2	Frankfurt	98.8
3	Milan	98.1
4	Brussels	96.9
5	Mexico City	96.0
6	Düsseldorf	94.6
7	Hong Kong	93.3
8	London	92.8
9	Sydney	88.4
10	Madrid	87.3
11	Amsterdam	83.7
12	Toronto	83.2
13	Sao Paulo	78.4
14	Tokyo	70.0
15	**Boston**	52.7
16	**LOS ANGELES**	51.8
17	**Dallas**	45.3
18	**NEW YORK**	44.7
19	**San Francisco**	42.6
20	**Chicago**	42.5
21	**DETROIT**	39.4
22	**Atlanta**	35.3
23	Seoul	33.9
24	**Minneapolis**	0.42
25	**Philadelphia**	0.00

Source: Taylor (2008).

Note
Interpretation: these figures are computed from *Advertising Age* data on the top nine companies listed in Table 5.4 in relation to data in Table 5.3.

completely dominated by smaller local firms that are not involved in global work. Philadelphia is emblematic of the depth of US advertising resulting from American exceptionalism but also the 'localness' of much of this advertising work.

It should not, however, be assumed that we, therefore, believe the dynamics of 'local' markets are not important in explaining the geography of advertising globalization. Each city market and its characteristics are, in fact, central to our explanation. For example, Table 4.6 shows US and non-US cities intermingled in terms of commission rates for advertising. Although there are agreed levels of commission across the industry, there remain supply and demand features in the deviations from the norm. This is shown by the top and bottom ranked cities in Table 4.6. Tokyo has its own complex of advertising firms, with relatively little penetration from outside, resulting in a sellers' market. In contrast, Detroit's prowess in advertising results from just one industry – motor manufacturing – with a few major purchasers: it is a classic buyers' market. Such factors are central to any explanation and are returned to in Chapters 6, 7 and 8.

Table 4.6 Estimated commission rates by city

Rank	*Cities*	*% Commission rates*
1	Tokyo	12.85
2	Düsseldorf	12.29
3	**San Francisco**	11.94
4	Seoul	11.86
5	Brussels	11.63
6	**Atlanta**	11.62
7	**Chicago**	11.54
8	**NEW YORK**	11.52
9	Toronto	11.48
10	**LOS ANGELES**	11.38
11	Amsterdam	11.38
12	Hong Kong	11.25
13	**Dallas**	11.18
14	Mexico City	11.12
15	Sao Paulo	11.08
16	Madrid	11.02
17	London	10.98
18	Paris	10.91
19	Frankfurt	10.88
20	**Boston**	10.72
21	Milan	10.63
22	Sydney	10.51
23	**DETROIT**	10.42

Source: Taylor (2008).

Note

Interpretation: these results are computed from data in *Advertising Age*.

Advertising in the early years of the twenty-first century

Unfortunately the *Advertising Age*'s data centre stopped collating information for cities in 2001 and, therefore, the city level findings above cannot be updated to cover the specific period of our interviews in 2007. However the GaWC global survey of advanced producer service offices is ongoing and we use this data for 2008 to provide an extensive context for our triple city study. It is particularly valuable that we can compare 2000 and 2008 results because general trends in advertising globalization in the early years of the twenty-first century can be charted. Without the *Advertising Age* material, we support the GaWC analyses for this time with data specifically on our case study cities drawn from the US Bureau of Labor Statistics.

Advertising agencies in world city networks – 2008

The GaWC data collection for 2008 was more comprehensive than for 2000: 175 advanced producer service firms were investigated and their offices evaluated across 525 cities. There were twenty-five advertising agencies included in the data.

Table 4.7 ranks cities in terms of global network connectivity in advertising in 2008. It is directly comparable with Table 4.1. After noting that this list repeats the domination of New York in advertising networks – its lead over London marginally increases from 2000 – the key difference between the two tables is that Table 4.7 is much longer. This is because in order to include our three case study cities we have to go outside the top 100: Detroit ranked a lowly 111th in 2008. This is not a singular service decline for Detroit; it is a general feature of relative decline for US cities in general. Across all advanced producer services, increasing globalization of the world economy has resulted in further diffusion of necessary services worldwide, leaving US cities relatively less important globally (see Taylor and Lang 2005; Taylor and Aranya 2008; Derudder *et al.* 2010). Thus even though Chicago's advertising connectivity has risen from 0.44 in 2000 (Table 4.1) to 0.47 in 2008 (Table 4.7), its world ranking has declined fifteen places from twenty-fifth to fortieth. However, specifically in advertising, global connectivities have generally declined absolutely from 2000 to 2008. For Los Angeles and Detroit major declines in global ranking are the result of large absolute declines: 0.07 for the former and 0.13 for the latter. Of course, our measurement method – scoring the top city 1.00 – means that we cannot directly gauge the changing status of New York since 2000 in Tables 4.1 and 4.7. If the 'imperial roles' of the New York offices of advertising agencies were being reduced, as suggested previously, then New York should exhibit some relative decline in global connectivity for this service sector. It is possible to show that this is indeed the case.

Figure 4.1 shows the cumulative decline in advertising connectivity by rank for the top 100 cities for both 2000 and 2008. The difference between the two curves is striking. After London (rank 2) the 2008 curve bulges out above the 2000 curve until they gradually converge again at rank 100. What this means is

Table 4.7 Global network connectivities (GNC) of cities in the advertising industry, 2008

Rank	*City*	*GNC*	*Rank*	*City*	*GNC*
1	**NEW YORK**	1.00	51	Ljubljana	0.39
2	London	0.75	52	**San Francisco**	0.39
3	Paris	0.75	53	Barcelona	0.39
4	Hong Kong	0.73	54	Kiev	0.38
5	Tokyo	0.71	55	Melbourne	0.38
6	Singapore	0.70	56	Tallin	0.38
7	Moscow	0.65	57	**LOS ANGELES**	0.37
8	Shanghai	0.64	58	Bangalore	0.37
9	Warsaw	0.63	59	Zagreb	0.37
10	Sydney	0.63	60	Beirut	0.37
11	Brussels	0.62	61	Ho Chi Minh City	0.37
12	Buenos Aires	0.62	62	Auckland	0.37
13	Taipei	0.61	63	Karachi	0.36
14	Mumbai	0.61	64	Frankfurt	0.36
15	Toronto	0.61	65	Cairo	0.35
16	Athens	0.60	66	Montevideo	0.35
17	Stockholm	0.60	67	Riga	0.35
18	Beijing	0.60	68	Oslo	0.35
19	Bangkok	0.60	69	Chennai	0.34
20	Madrid	0.60	70	Lima	0.33
21	Milan	0.60	71	Casablanca	0.33
22	Seoul	0.59	72	Santo Domingo	0.33
23	Budapest	0.57	73	Panama City	0.33
24	Vienna	0.56	74	Vilnius	0.32
25	Istanbul	0.56	75	Guatemala City	0.32
26	Kuala Lumpur	0.55	76	Belgrade	0.32
27	Helsinki	0.55	77	Cape Town	0.32
28	Dubai	0.54	78	**Atlanta**	0.32
29	Lisbon	0.54	79	Bogota	0.31
30	Mexico City	0.53	80	Nairobi	0.31
31	Amsterdam	0.53	81	Bratislava	0.31
32	Jeddah	0.53	82	Guangzhou	0.30
33	Copenhagen	0.52	83	**Miami**	0.29
34	Bucharest	0.52	84	Manama	0.29
35	Rome	0.51	85	San Salvador	0.28
36	Prague	0.51	86	Colombo	0.28
37	Caracas	0.50	87	Düsseldorf	0.28
38	Dublin	0.49	88	Hamburg	0.28
39	Sao Paulo	0.47	89	Almaty	0.27
40	**Chicago**	0.47	90	**San Jose (CA)**	0.25
41	Jakarta	0.46	91	Nicosia	0.25
42	Zurich	0.46	92	San Juan	0.25
43	Johannesburg	0.44	93	Denver	0.25
44	Kuwait	0.44	94	Skopje	0.24
45	New Delhi	0.44	95	Amman	0.24
46	Sofia	0.42	96	Guayaquil	0.24
47	Tel Aviv	0.42	97	Managua	0.24
48	Manila	0.41	98	Berlin	0.23
49	Riyadh	0.41	99	Tegucigalpa	0.23
50	Santiago	0.41	100	**Minneapolis**	0.22

Table 4.7 continued

Rank	*City*	*GNC*	*Rank*	*City*	*GNC*
101	Accra	0.22	111	**DETROIT**	0.18
102	Quito	0.22	112	Minsk	0.18
103	Rio de Janeiro	0.22	113	Harare	0.18
104	Munich	0.22	114	Algiers	0.17
105	Lagos	0.22	115	Calcutta	0.17
106	Sarajevo	0.20	116	Asunción	0.17
107	Dhaka	0.20	117	Lahore	0.16
108	**Washington**	0.20	118	Islamabad	0.15
109	**Dallas**	0.19	119	Abidjan	0.15
110	Geneva	0.18	120	**Seattle**	0.15

Source: authors' research and GaWC data (see www.lboro.ac.uk/gawc (accessed 1 September 2010)).

Note
For interpretation see Table 4.1.

that the cities with major and medium concentrations of advertising below New York and London have augmented their advertising connectivities to a far greater degree than the top two cities. Specifically this shows relative decline of 'imperial' New York. The bulge in Figure 4.1 can be defined as the sequence between where increases of 0.10 begin and end. These cities are listed in Table 4.8. Apart from the special case of China with its three leading cities featured, this list identifies the leading cities in medium-sized consumer markets across all continents. This can be interpreted as initial evidence for the breakdown of New York imperial relations in advertising.

In reporting the city office strategies for 2000 we had to use an analysis of all advanced producer firms because of the relatively low number of advertising agencies in the data (seventeen). In that analysis we were able to find the advertising imperial strategy from among the overall office location strategies (Table 4.2). For 2008, with twenty-five advertising agencies in the data, we can conduct a principal components analysis featuring just these twenty-five arrayed across the top 140 cities. Using this technique three main advertising office strategies are revealed that account for just under 50 per cent of the variation in the data. That is to say the twenty-five individual firms' strategies can be reduced to just three common patternings of offices. The three common strategies will be described in order of their statistical importance.

Given the overall size of its market, it is not surprising that the most important component shows a European focused strategy (Table 4.9). With London and New York heading the list, there is an inter-weaving of eastern and western European cities plus a few non-European cities with strong links to the region. There are twelve agencies that feature in constituting this common pattern through their office networks.

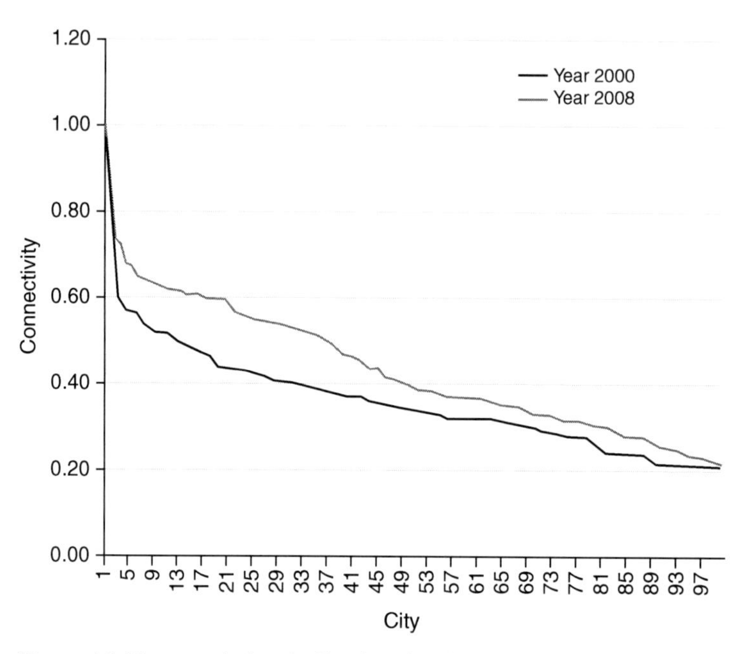

Figure 4.1 The cumulative decline in advertising connectivity by rank for the top 100 cities for both 2000 and 2008 (source: authors' research and GaWC data (see www.lboro.ac.uk/gawc (accessed 1 September 2010))).

Note
Key to cities represented by numbers on x-axis provided in Table 4.7.

Table 4.8 Change 2000–08: cities of the bulge shown in Figure 4.1

Paris	Budapest
Hong Kong	Vienna
Tokyo	Istanbul
Singapore	Kuala Lumpur
Moscow	Helsinki
Shanghai	Dubai
Warsaw	Milan
Sydney	Lisbon
Brussels	Mexico City
Buenos Aires	Amsterdam
Taipei	Jeddah
Mumbai	Copenhagen
Athens	Bucharest
Toronto	Rome
Stockholm	Prague
Bangkok	Caracas
Beijing	Dublin
Madrid	Sao Paulo
Seoul	Chicago

Source: authors' research and GaWC data (see www.lboro.ac.uk/gawc (accessed 1 September 2010)).

Table 4.9 European focus strategy (18.71 per cent)

City	*Score*	*Agency*	*Loading*
London	2.68	ZenithOptimedia	0.70
NEW YORK	2.23	McCann Erickson Worldwide	0.68
Budapest	2.16	Euro RSCG Worldwide	0.67
Warsaw	2.03	Mindshare Worldwide	0.65
Brussels	2.01	Com Vort Group	0.61
Milan	1.90	Publicis	0.57
Moscow	1.84	Young & Rubicon	0.57
Stockholm	1.83	BBDO Worldwide	0.54
Rome	1.75	Wunderman	0.52
Prague	1.74	Starcom MediaVest Group	0.48
Lisbon	1.69	JWT	0.42
Dublin	1.69	Grey Worldwide	0.40
Madrid	1.64		
Vienna	1.64		
Oslo	1.61		
Casablanca	1.51		
Dubai	1.40		
Barcelona	1.39		
Bucharest	1.27		
Buenos Aires	1.22		
Athens	1.15		
Helsinki	1.12		
Sydney	1.04		
Paris	1.03		

Source: authors' research and GaWC data (see www.lboro.ac.uk/gawc (accessed 1 September 2010)).

Note

Interpretation: these results are from a principal components analysis of twenty-five advertising agencies across 140 cities using a varimax rotation. The scores are as in Table 5.2 and all cities scoring above 1.0 are listed. The loadings on agencies provide a measure of the contribution of an agency to the constitution of the common office location strategy. Component loadings have properties of correlation coefficients and vary between +1 and –1. Agencies with loading above +0.4 are shown. The percentage bracketed with the title indicates how important the component is within the overall analysis; it measures the percentage of all the variation in the data that is included in the component.

It is the second component that is particularly interesting for our general purposes here. This is shown in Table 4.10 as a New York articulated strategy that is similar to the imperial advertising strategy identified for 2000 in Table 4.2. Does this mean that the imperial strategy survives? Maybe, but the differences between the two tables imply at least an amelioration of imperial relations. For a start, the degree of New York's dominance is lessened by about 20 per cent (from a score of 3.2 to 2.6). Furthermore, the scores for the other cities are raised rank for rank. This replicates the finding of Figure 4.1 with its 'bulge cities'. In fact, nine of the fourteen cities below New York in Table 4.10 are bulge cities in Table 4.8. The main agencies whose offices constitute this common pattern are

Table 4.10 New York international strategy (16.13%)

City	*Score*	*Agency*	*Loading*
NEW YORK	2.63	Ogilvy and Mather Worldwide	0.80
Caracas	1.84	OgilvyOne Worldwide	0.76
Sao Paulo	1.78	Leo Burnett Worldwide	0.67
Mumbai	1.63	Starcom MediaVest Group	0.53
Tel Aviv	1.51	Grey Worldwide	0.52
Amsterdam	1.49	JWT	0.51
Nairobi	1.49	DraftFCB	0.50
Copenhagen	1.39	BBDO Worldwide	0.46
Toronto	1.38	Satchi & Satchi	0.46
Johannesburg	1.31	TBWA Worldwide	0.45
Chicago	1.26		
Colombo	1.18		
Istanbul	1.17		
Athens	1.15		
Kuwait	1.04		

Source: authors' research and GaWC data (see www.lboro.ac.uk/gawc (accessed 1 September 2010)).

Note
For interpretation see Table 4.9.

also listed in Table 4.10: the results suggest these firms are working less in a traditional New York imperial mode.

The third advertising location strategy has a distinctive Pacific Asian focus (Table 4.11). Articulated mainly through Tokyo (with the highest score in the analysis) and Hong Kong, this strategy features mainly Pacific Asian cities plus other cities with strong relations to the region (Indian cities, Sydney and Dubai). Although statistically important, it is mainly constituted by fewer agencies – just five – than the other components. The importance of Paris is a surprise here but the interesting result from our project's perspective is the strength of Los Angeles.

Following on from the latter point, Table 4.12 shows all the scores for the three case study cities in the principal components analysis. As well as New York's ubiquitous status and Los Angeles' Pacific Asian link documented in previous tables, these results show negative scores. That is to say, Los Angeles' mix of agency offices is largely made up by firms without either a European or New York international strategy. Detroit's advertising office mix is even more detached: it combines firms without the three main global strategies we have identified. In other words, Detroit completely, and Los Angeles mainly, are part of the 50 per cent of office strategies that have particular rather than common global patterning. As so, we return as before to American exceptionalism, albeit with the massive exemption of New York in advertising world city networks.

Table 4.11 Pacific Asian focus strategy (12.93 per cent)

City	*Score*	*Agency*	*Loading*
Tokyo	4.29	Hakuhodo	0.84
Hong Kong	3.32	Asatu/DK	0.83
Paris	3.03	Densu	0.55
Singapore	2.86	ODM	0.54
Beijing	2.79	TBWA Worldwide	0.49
Bangkok	2.46		
Seoul	2.30		
Kuala Lumpur	2.11		
New Delhi	2.10		
Shanghai	2.01		
Osaka	1.48		
Ho Chi Minh City	1.40		
Guangzhou	1.31		
Mumbai	1.29		
Jakarta	1.25		
LOS ANGELES	1.23		
Düsseldorf	1.17		
Sydney	1.12		
NEW YORK	1.10		
Dubai	1.10		

Source: authors' research and GaWC data (see www.lboro.ac.uk/gawc (accessed 1 September 2010)).

Note
For interpretation see Table 4.9.

Table 4.12 Case study cities within the location strategies

City	*European focus*	*New York international*	*Pacific Asian focus*
New York	**2.23**	**2.63**	**1.10**
Los Angeles	–0.07	–0.56	**1.23**
Detroit	–0.93	–0.19	–0.47

Source: authors' research and GaWC data (see www.lboro.ac.uk/gawc (accessed 1 September 2010)).

Notes
Emboldened scores are previously recorded in Tables 4.9, 4.10 and 4.11
Interpretation: component scores are standardized, which means that they have a mean of zero and are given in units of standard deviation. Thus scores in previous tables show important cities defined as having scores above one standard deviation greater than the mean (0). In this table there are measures less than the mean, which are shown as negative values.

Advertising agencies and employment in New York, Los Angeles and Detroit

Finally we provide some basic contextual statistics on our three case study cities. The US Employment Census is used for obtaining data on the metropolitan areas encompassing our three cities in the years leading up to the completion of interviews. As noted, more data is provided in later chapters in the book. The aim here is to set the scene for this discussion.

Table 4.13 indicates the relative sizes of the advertising sector in each city as represented by number of agencies. As expected New York has by far the most agencies – about twice as many as Los Angeles – and Detroit has by far the fewest. For the former two cities we can show data on number of employees and the payroll. The interesting feature here is to be found in the ratio between these two figures that shows higher remuneration in New York, presumably reflecting the importance of its agency offices in the advertising world market.

The low number of agencies in Detroit is, perhaps, a surprise given the city's relative prominence in US advertising – see Table 4.3. Detroit is a large market served by a limited number of agencies. This can be partly explicated by the

Table 4.13 Advertising agencies in New York, Los Angeles and Detroit, 2004

Metropolitan area	*Number of agencies*	*Number of employees*	*Payroll (first quarter)*	*Ratio (employees to payroll)*
NEW YORK*	1,508	28,780	610,825	21.22
LOS ANGELES**	776	10,716	191,906	17.91
DETROIT***	181	N/A	N/A	N/A

Source: US Census Bureau 2007.

Notes
* New York–Newark–Edison Metropolitan Statistical Area.
** Los Angeles–Long Beach–Santa Ana Metropolitan Statistical Area.
*** Detroit–Warren–Livonia Metropolitan Statistical Area.

Table 4.14 Distribution of advertising agencies by size across metropolitan areas, 2004

*Metropolitan area**		*Number of employees*			
		1 to 9	*10 to 99*	*100 to 499*	*above 500*
New York	Total	1,122	331	49	6
		74.35%	21.94%	3.25%	0.40%
Los Angeles	Total	584	172	19	1
		75.26%	22.16%	2.45%	0.13%
Detroit	Total	138	34	4	5
		76.24%	18.78%	2.21%	2.76%

Source: US Census Bureau 2007.

Note
* For definitions, see Table 4.13.

analysis in Table 4.14. This shows the distribution of agency sizes across our three cities. Distributions are fairly similar apart from one large difference. Detroit has proportionally many more agencies in the largest category than New York and Los Angeles. In fact, despite having only about one-eighth of New York's number of agencies, Detroit is nearly on a par with the larger city for agencies employing over 500: there are six of these large agencies in New York to five in Detroit. Clearly Detroit is compensating to some extent for its relative dearth of agencies with a concentration of very large agencies. This is a function of the city's advertising market dominated by the automobile industry as noted previously. As the home of a key industry in US advertising demand, Detroit attracts a few very large agencies to provide the necessary supply at source.

Implications and conclusions

In this chapter we have drawn on a range of quantitative analyses to set the scene for subsequent discussions of the geographies of advertising globalization, the work processes of global agencies and the role of our three case study cities (New York, Los Angeles and Detroit) in these geographies and work processes. There are two main implications of the discussion.

First, it is clear from analyses that a combination of territorial and network assets defines the roles of different cities in advertising globalization and advertising work processes. In the case of the exceptional USA this relates to the differing roles of cities in advertising globalization, roles which range from intense involvement (for example, New York) to moderate involvement (for example, Los Angeles) and even exclusion (for example, Philadelphia). Similar variations in levels of involvement can be noted for cities worldwide. Second, the analysis also reveals that the processes of advertising globalization themselves have important geographies. Specifically it has been shown that the 'imperial' New York outwards strategy has evolved as Europe and Pacific Asian strategies have led to differing roles for cities, as exemplified by Los Angeles' important role in Pacific Asian strategies, but relatively unimportant role in European strategies.

Building on these insights, the rest of the book proceeds to explore the way work is organized in global agencies and the implications of organizing strategies for the role of different cities in agency networks (Chapter 5), and the way such organizational strategies when combined with the territorial assets of cities determine the city geographies of advertising globalization (Chapters 6, 7 and 8).

5 Agencies and advertising globalization

Coordinating interactions with clients and consumers

To build on discussions in Chapter 4 and continue the task of analysing the geography of the work of global advertising agencies, this chapter draws on data collected through interviews with advertising executives to unpack the spatiality of the different management, strategic and creative tasks associated with the development of advertising campaigns. This unpacking reveals the multiple and inter-twined geographies of the three different types of advertising work, and shows how cities pin down each type of work in different ways depending on their territorial assets and network connectivity. In particular, it is shown that cities can capture advertising work by manufacturing themselves a role in client relationship management, strategic planning or creative work. Those cities that play a role in all three stages of work, management, planning and creativity, and that attract and generate global flows of such work, become the most successful advertising centres. The analysis, therefore, provides a way of further exploring the different roles and network connectivities of cities identified in Chapter 4 and sets the strategic context for subsequent chapters on New York, Los Angeles and Detroit and the role of these different US cities in the work of global advertising agencies.

Global agency work: divisions of labour between account managers, planners and creatives

As knowledge-intensive business services, advertising agencies are reliant on their workers and their embodied and encultured knowledge because of their role in the problem solving associated with campaign development (Blackler *et al.* 1998; Hackley 1999; Nachum 1999). Consequently, it is unsurprising that the agencies studied as part of this research were staffed by highly trained and skilled workers holding, for example, degrees including MBAs and PhDs awarded by universities from throughout the world in cities such as Mexico City, Singapore, London, Brasilia, Beijing and Shanghai, as well as several US cities. With an average churn rate of between 20 and 30 per cent per annum being reported as workers moved between agencies, recruiting such talented workers was, in 2007, a major challenge for all of the agencies we studied. This challenge was exaggerated by the need of advertising agencies to recruit a range of workers with a skills portfolio that is perhaps more diverse than in any other comparable knowledge-intensive business service sector.

Whilst it has been widely reported that the emergence of 'managed professional business' has led to knowledge-intensive business service firms such as architecture, accountancy and law employing not only members of the profession (i.e. architects, accountants and lawyers), but also a range of other 'experts' such as professional finance and human resource managers, such diversity in the employee-base is usually associated with the development of a management cadre rather than the enhancement of the skills of those involved in the service production process (see for example, Cooper *et al.* 1996; Brock *et al.* 1999). Finance and human resource managers ensure the smooth running of the firm, leaving the production and delivery of services to a relatively homogeneous group of, for example, architects, accountants or lawyers, distinguished by their shared educational credentials in the shape of a common university degree or professional qualification (for example, in architecture, accountancy or law) (Faulconbridge and Muzio 2008). In the case of global advertising agencies, the management jobs associated with the smooth running of the organization are predominantly handled at the holding group level with a small proportion of an agency's workforce having titles such as Chief Executive Officer or Managing Partner and reporting to holding group executives. However, this does not mean agencies are populated by a homogeneous group of individuals with shared educational credentials, skills sets and abilities. Rather agencies are made up of three groups of workers – account managers, account planners and creatives – that are all intimately involved in the production process, but have very different skill sets and abilities and come from very different educational backgrounds.

Account managers are primarily responsible for an agency's relationship with clients who commission advertising campaigns. Account management personnel are charged with ensuring the client's aims and demands are met by those producing an advert. As one interviewee commented, the importance of this role cannot be underestimated because, '[c]lients want to know that you care. They wanna know that their priorities matter to you, that you're engaged with trying to solve the problem' (Interviewee 7). The skills of an account manager are very much associated with client relationship management, but, at the same time, also require individuals to be competent project managers capable of ensuring the two other groups of workers involved in developing advertising campaigns, account planners and creatives, fulfil the client's needs within the set timescales and budgets. Account managers have an ongoing role throughout a campaign and are the coordinators of all of the activities involved in developing an effective end product for television, radio, the printed press or, increasingly, the internet. Many account managers are often ex-planners who have moved into the account management role after demonstrating their client relationship management skills or have completed management related degrees.

Account planners were, in the early stages of agency development, part of the creative team associated with a campaign. However, as part of second and third wave advertising and responses to post-Fordist consumption cultures and reflexive, individualistic and spatially entangled consumers, a separate account planner's role was developed by a number of leading agencies and has now become an institutionalized position within all agencies (Mattelart 1991). Planners are responsible for translating the client's aspirations, both in terms of brand development and campaign outcomes, into a strategy that can be deployed in an advertising campaign. This includes identifying target audiences and developing strategies to tap into consumers' identities, emotions and desires as part of a campaign. Account planners now fulfil a central role in the development of a successful advertising campaign. As one interviewee described the skill-set needed in relation to planning:

> We have a team of planners. Currently, there are about 15 planners, real planners at [firm x]. Then I have another group who work with planning called 'marketing science'. They're actually all women, they're all analytics PhDs. So they take existing data that we have on markets and they do what I would describe as the 'heavy lifting' of analytics. And then, we have what we call 'cultural knowledge'. And there are two aspects to cultural knowledge. One is trends and analysis. And then we have a full-time PhD anthropologist.
>
> (Interviewee 4)

As the quotation suggests, the skills of an account planner are very different to that of an account manager, with university-level training providing the basis for some of the types of analytical work that have to be completed. Indeed, it is the account planner's role that is most likely to be filled by an individual with a marketing related degree with other roles requiring skill sets not necessarily directly associated with marketing. In terms of day-to-day work, at one level the planner must be able to collaborate with the account manager to ensure that the client's needs are fulfilled. This also means having the skill and ability to interact with client themselves, extrapolate an understanding of their brand, product and marketing needs and convince them of the merits of any strategy developed by the agency. At another level, the account planner also has a 'translation' role in that they must also be able to effectively find a way to turn any research and analysis completed into an advert for whatever medium the client wishes to use or the agency has advised the client to use. To do this they must interact with the agency's creatives.

Creatives are responsible for developing the storyline, artwork and visual medium upon which an advert relies. They are also responsible for the execution of an advert in terms of relationships with studios, artists, producers etc. As such the creative role is often, but not exclusively, filled by individuals with art or media related degrees (for example, fine or contemporary arts, film studies, design). Working with the account planners, creatives will seek to turn a 'rational' strategy into an advert that irrational, emotional and reflexive consumers will respond to, and, most importantly, respond to in a way that creates demand for a client's products. As a result, agencies are always on the lookout for creative individuals who have both the ability to develop innovative campaigns but also the ability to manage the execution of creative ideas through the exploitation of the project ecology described in Chapter 3. Indeed, in a highly competitive industry, an advertising agency's reputation, market position, profitability and ability to enter new geographical markets is very much reliant on the credentials, inspiration, ideas and abilities of its creatives because, in the eyes of clients or potential clients, it is the creative product of agencies and its effects on consumers, rather than the underlying strategy and its sophistication, that is most critically assessed.

Accordingly, creative workers play a major role in the genesis and operation of all advertising and creative jobs and are, therefore, often the ones that add the most value to a campaign. Without a successful team of creative workers that endeavour to stay one step ahead of the competition so as to enable campaigns to respond to changes in consumer trends through innovative advertising designs, an agency is likely to find it difficult to compete in the cut-throat advertising marketplace. Hence agencies view their creatives as, 'the most valuable and the most expensive people … [who] … represent probably a third of the people … [in the agency] … but, 50 to 60 percent of the payroll' (Interviewee 5). Indeed, in a market which has increasingly been plagued by concerns about the diminished effectiveness of advertising, and in particular the growing ineffectiveness of television based advertising as it becomes more and more difficult to attract the attention of consumers in an advertising saturated world, the importance of creative workers has grown significantly. Hence there is a relatively high premium placed on 'young' talented, creative and 'blue sky' thinking workers from an array of graduate and non-graduate sources. It is not surprising, therefore, that agencies invest heavily in the retention of creative staff by paying high salaries and permitting the open and unrestrictive working patterns that characterize the high degrees of autonomy associated with knowledge-intensive and professional work (Alvesson 2004; Faulconbridge 2008).

Advertising knowledge-bases: diversity and interaction

Our analysis of account planners, account managers and creatives in advertising reveals a key feature of the contemporary advertising agency: agencies have to manage a diverse pool of workers that all play an integral role in the advertising production process, but have different educational backgrounds and work practices and are drawn from different labour markets. In the context of the conceptual framing developed in Chapters 2 and 3 this is significant for two reasons. First, the advertising production process involves workers with multiple knowledge-bases. Whilst what Asheim *et al.* (2007) would call 'symbolic' knowledge is important in the work of creatives, advertising cannot be classified as solely an industry reliant on such knowledge. 'Synthetic' work by account planners is equally important. The account management role is harder to map on to the framework of Asheim *et al.* (2007), something which further highlights the danger of assuming that the knowledge-base of particular industries or jobs can be pigeon-holed into one category or another, and is perhaps best described as being a social process allied to the debates about the role of face-to-face contact. Table 5.1 captures these multiple knowledge-bases and forms of expertise that cut across the advertising production process. The reliance on such diverse forms of expertise has geographical implications with each job role in global agencies lending itself to different geographies of innovation (Table 5.1). As a result, territorial (city) and global (intra-agency) networks play different roles in relation to different types of work, creating multiple geographies of work in global agencies that are best understood as topological and defined by both territorially and network embedded forms of work (see for example, Hess 2004). This has implications for how we understand the role of cities in advertising globalization, suggesting the most successful advertising cities are those that somehow bring together all three types of advertising work and their associated topological embeddings. The remaining sections of the chapter, therefore, examine in detail the spatiality of the tasks of each of the three core groups of workers in agencies and how this defines the way agencies operate in and through cities. This provides insights that help explain the uneven and diverse geographies of advertising work in the twenty-first century that were mapped in Chapter 4.

Account management

The first job of the account manager is to understand exactly how the client wants the agency to manage a campaign. Of particular relevance here is the decision about whether to run a 'global' campaign or a coordinated and integrated 'transnational' campaign. As outlined in Chapter 2, the two are subtly

Table 5.1 The knowledge-bases associated with different advertising job roles

Job role	*Knowledge-base*	*Spatial implications for innovation*
Account management	Reliant on ability to interact effectively with clients, account planners and creatives to manage the production process	Proximity and face-to-face contact important to develop trusting relationship but virtual interactions with fellow agency colleagues likely to also be important
Account planning	'Synthetic' and 'symbolic' knowledge associated with planning strategies	Knowledge tends to be situated and whilst some insights can be diffused and developed at a distance others are tacit, encultured and embrained in individuals
Creatives	'Symbolic' knowledge dominates and used to connect planning strategies to campaign artwork, actor scripts etc.	Knowledge entirely encultured and embrained in individuals with diffusion difficult

different. 'Global' describes an approach in which an advert is used with limited adaptation in multiple markets (for example, adaptation in the form of language translation into French, German and Spanish from English). In reality such 'global' approaches are, however, often supra-regional (for example, EU-wide). This reflects the fact that many TNC clients tend to manage marketing budgets and strategies by supra-regional market, such as the Americas, Europe, the Middle East, Asia and Australasia. Transnational campaigns involve developing market-specific advertising, although markets do not correspond with one scalar fix, such as the 'local'. Some campaigns may be national, other regional, particularly within large countries such as the USA and Australia, whilst few are actually 'local', for example being designed for and run in one city only. Target markets or audiences are defined by socially constructed groups that, despite being smaller and smaller in size, share some basic commonalities in terms of consumption habits, emotional responsiveness and reflexivity. Such groups might exist within, but also across national, regional or local boundaries. Hence from hereon in when discussing transnational advertising and the development of multiple adverts for a client by multiple offices in an agency's global network, we primarily refer to market-specific campaigns to avoid scalar lexicons that indicate a direct relationship between agency strategies and nested, hierarchical scales.

Despite recognition that the 'global' campaign is increasingly inappropriate, when clients demand such a strategy, often because of financial constraints on marketing budgets which prevent investments in campaigns designed for one market only, an agency has no choice but to deliver such a product. As a result, agencies have to be capable of delivering multiple models of campaign. One account manager summarized the type of work he managed as follows:

> Today clients are either highly decentralised or fully globalised or there's a hybrid in between. [Client w] actually is very centralized but they do a lot of their activation – which is promotion type work – locally so they're sort of a hybrid. [Client x], they'll do a lot of their retail work locally … [Client y], there's templates that can be implemented locally so it's all globally done but then it's implemented locally. [Client z] remains totally decentralized, everything is created locally.
>
> (Interviewee 8)

It is the account manager's job to deliver such multiple-mode advertising that caters to client demands for either 'global' or 'transnational' market-specific campaigns through the effective management of the resources of the agency. As one interviewee described this role:

> They [clients] want one point of contact, the ability to have someone orchestrate a single point of contact for the company. And then the critical factor becomes how do you align the [advertising] organization, that group of

> people to work together, to collaborate in a way that's gonna have one outcome for the client. It's not always easy.
>
> (Interviewee 3)

There are two implications of this role of the account manager for the geography of account management work itself. First, account management work has its geographies defined by the geographies of clients. Second, account management work also has its geographies defined by the geographies of campaign management, with both 'global' and 'transnational' campaigns requiring account managers to work with colleagues in different offices of the advertising agency in different ways. Both the geography of client–account manager interactions and account manager–agency colleague interactions are unpicked further below.

Geographies of client and campaign management

There was general accord amongst advertisers that the client prefers to have a major office of the agency it employs close to their marketing headquarters and that this also brings benefits for the agency. As one interviewee suggested:

> They're [clients] not here because we're here. We're here because they're here. If they were in Sub-Sahara in Africa, I guarantee you there would be an agency in Sub-Sahara Africa and you would be interviewing and say 'what is it about Sub-Sahara Africa that attracts so many of you people?' The answer would be 'client's proximity'.
>
> (Interviewee 16)

Client management teams tend to be centralized in a 'lead office' which is in proximity to the client's marketing manager and their team. This mirrors the ideas encapsulated in theories of agglomeration economies with face-to-face contact with the client being critical for developing an effective business relationship (Daniels 1993; Jones 2007; Pratt 2006). At one level, then, advertising is affected by 'compulsions of proximity' (Goffman 1967; Boden and Molotch 1994) that have long been tied to the value of embodied interactions during communication, problem solving and rapport and trust development. Account managers constantly struggle to keep clients happy, to understand their marketing needs and convince them of the value and appropriateness of the agency's work. Being able to meet face-to-face regularly without travelling long distances is crucial. As one interviewee noted, 'I think that personal relationships are really good and productive with clients and the opportunity to physically talk face to face affords you possibilities and affords you the opportunities to submit relationships that aren't available through teleconferences' (Interviewee 16).

It is unsurprising, therefore, that historically offices in cities such as New York played a central role in the work of global agencies. As agglomerations of clients and in particular the headquarters of TNCs (Sassen 2000), the high demand for advertising work in major world cities quickly led to offices in cities such as New York becoming powerful and central nodes in the global networks of agencies. Most recently, and as part of the third wave advertising described previously, the importance of such offices and their proximity to clients has grown further because of an increasing demand for advertising that responds *quickly* to changes in consumer behaviours. As part of attempts to develop advertising tailored to the current mood of consumers, moods influenced for example by major sports or political events, timescales for the development of campaigns have been shortened. Advertisers are increasingly being asked to deliver presentations of campaign plans within weeks of the completion of a pitch, and sometimes as little as two weeks, instead of within months. This reinforces the importance of proximity between account managers and clients: proximity facilitates regular interactions, often at short notice, between account managers and clients when campaigns are developed in short timescales.

At first glance and in line with existing literature that documents the role of agglomeration economies, account management tasks would seem to be associated with the reinforcement of a geography of advertising dominated by what in the GaWC typology would be called Alpha world cities (Beaverstock *et al.* 1999b and www.lboro.ac.uk/gawc/gawcworlds.html (accessed 1 September 2010). Specifically, it might be expected that cities such as Paris, New York, Los Angeles and London would continue to be the powerhouses of advertising work in the twenty-first century because of the agglomerations of clients and in particular major TNCs that exist there. To a certain extent this is true. As Chapter 4 revealed, in many ways the geographies of advertising work continue to reflect the geographies of US and Western European TNCs and their headquarters. However, account management work has also been developing important new geographies over the past twenty years which have brought what in GaWC typologies are called Beta and Gamma world cities, such as Bangkok, centre stage in the geographies of advertising. There are three important forms of change in the work of account managers that, alongside changes in account planning and creative work which are reviewed below, lie behind such developments.

First, many clients are increasingly decentralizing marketing activities away from headquarters and fragmenting responsibility for marketing for different products across the organization. In line with the shift towards the transnational form of firm management that Bartlett and Ghoshal (1998) describe, this means budgets for marketing and personnel charged with managing these budgets are located in multiple locations both within one country (for example multiple locations within the USA) and internationally within client organizations. As one advertiser noted, this can also mean 'a lot of companies have moved to having an office doing their advertising, hiring another one to do their direct mail and another one to do their online' (Interviewee 20). This creates a new challenge for global agencies: the need to interact with commissioning clients in multiple places and not just provide post boxes that can deliver advertising commissioned in Alpha world cities. As a result, the number of offices with a role in managing agency–client interactions has increased significantly. No longer do the Alpha world cities hold all of an agency's key client relationships. Instead, account management duties are much more geographically dispersed.

Second, and reinforcing the trend towards the decentralization of client management work, the past ten years in particular has also seen clients originating from outside North America, Western Europe and Japan gain increasing strategic importance for global agencies in terms of revenues. Reflecting a further move away from what Mattelart (1991) describes as the 'imperialist' export model in which the global offices of advertising agencies provide services to US and Western European clients, global agencies are increasingly reliant upon the business of clients, and TNC clients, originating from South America, China and South East Asia. As a result, account management work has grown in importance in, for example, the Rio de Janeiro, Beijing, Shanghai, Singapore and Bangkok offices. As one advertiser noted, '[n]ew markets are opening up and creating incremental opportunities for agencies. So obviously that's how we wanna grow our network globally' (Interviewee 2). Further capturing this trend, two advertisers in Los Angeles provided examples of the changes afoot:

> Typically, the Singapore office would wait for the LA office to have a project. And here's something [shows interviewer poster] where the Singapore office led it, drove it and brought us in as needed. They're gonna own that piece of business and we're gonna serve them and I think that's a much more virtuous model.
>
> (Interviewee 22)

> In Beijing, we just won the business, we only have about ten people there. And in China the market's grown so quickly. When they did their pitch, we developed two or three campaigns from here to help them; we sent three of our account people over there to help with the pitch. And actually afterwards two of them moved to China from this office.
>
> (Interviewee 28)

Third, the geography of account management work has also been affected by clients reassessing the importance of proximity in decisions about which agency to use for a campaign. Some clients have begun to question whether agencies located in proximity to their marketing managers can always provide the best services and have begun to make compromises in terms of ease of 'meetingness' (Urry 2004) in order to tap into expert labour pools located at a distance that will deliver the most successful advertising campaigns. As one Detroit based interviewee noted:

> [Motor manufacturer x] did all their advertising out of state with a company in Boston and now it's with Crispin in Miami. So they've kind of broken the model. [Motor manufacturer y] recently did it as well. They shifted business to [agency x] in New York. They're starting to realize that maybe they can take parts of the business and send it out for a fresh perspective.
>
> (Interviewee 20)

Whilst the ideas outlined in the earlier discussion of the value of proximity are not rendered redundant by such developments – this trend towards choosing agencies located at a distance remains an atypical approach – the fact that clients are seeking out agencies located at distance that are perceived to offer performance advantages does highlight a subtle change in client behaviours that has implications for the geography of advertising work and account management work. In particular this trend means that whilst, at one level, the geographies of account management work continue to reinforce existing geographies and the dominance of cities such as New York and London because of their critical mass of clients, at another level it is only the cities that are also able to attract network flows of client revenues – i.e. the purchasing of services by clients located outside the city – that are truly successful.

In addition, the account managers' intra-agency role in overseeing the progression of a campaign as part of their responsibility for ensuring clients needs are met has also developed new spatialities. As already noted, the growing importance of the 'transnational' campaign has reconfigured the way agencies work to deliver multi-market advertising. For account managers this creates a new challenge: the need to choreograph the work of account planners and creatives in multiple offices. Account managers need to ensure those working on campaigns in other offices meet deadlines and remain within set budgets and, most importantly, must ensure, with the help of account planners and creatives, that the client's branding strategy is consistently communicated through adverts developed for different markets. This means the account manager that holds the client relationship – i.e. handles interactions with the client's marketing manager – must engage in transnational collaboration with fellow account management workers but also members of the account planning and creative teams in different offices. As a result, it would be misleading to suggest that forms of power relation are completely evacuated in the transnational form. Existing work on, for example, legal TNCs shows that the transnational form involves influential individuals, often from the largest office of the firm, constructing themselves resources through which they can exercise domination or coercion of groups in other offices (see for example, Faulconbridge 2007b, 2008; Jones 2007). Similarly in retail TNCs it has been shown that the transnational model involves two-way learning with headquarters learning from innovations developed in subsidiaries but also the imposition of best practices as part of firm-wide coordination strategies (Currah and Wrigley 2004). As such it is the way that individuals and offices develop and exploit resources that determines the geographies of power in global agencies. So as one agency manager outlined in relation to a global account for a major motor manufacturer and the power relations constructed between offices working on the account:

> the decisions are still made on the local ground level, whether it be in Zurich for Europe or Shanghai for China. That's where the decisions are being made, the money is being spent, the resources deployed. But we share ideas and people much more. If we have a challenge of a new product being

> launched, we will often tie-in now since with email, teleconferencing, everything else, you can send a brief around the world and easily get feedback, ideas without much of a problem as opposed to when I started, it was all by fax. You didn't pick up the phone very much because it was expensive to call overseas. Now it's just easy with what the Internet's done to the world. It's so easy to transfer information and have everybody be up to speed.
>
> (Interviewee 18)

The transnational collaboration described in this quotation is examined further in later sections of this chapter. In terms of account management such developments are significant because of the changing nature of the power relations between offices associated with a 'transnational' approach to advertising. Account managers that directly liaise with the client must maintain a delicate balance between collaboration with colleagues in other offices and control of the work produced. As one interviewee commented:

> there needs to be a clear direction coming from someone. It has to be the decision-maker. On the other hand, in this case, on this pitch, I was the decision-maker. That decision-maker as much as he needs to take control, I don't think the world works when everyone's got a shared opinion completely. Someone needs to lead it and have a point of view. But I do think that leader needs to listen. It's a strange thing to say but I would call it a 'very soft dictatorship'.
>
> (Interviewee 6)

Such a 'soft dictatorship' is vital if the client is to feel that the account manager has control of the overall advertising package being delivered by the agency. As one interviewee noted, 'the last thing a really beleaguered and busy client wants to manage is multiple offices and multiple motivations' (Interviewee 3). However, as already noted, the geography of account management work, and specifically the geography of account managers who handle the agency's relationships with clients, has changed over recent years and, as a result, 'soft dictators' now operate from multiple offices in an agency's network. Consequently it is not just a few select offices such as those in London or New York that have the potential to act as the 'command and control points' (Sassen 2000) for advertising work. Reflecting the assertion of Jones (2002) that hierarchical rankings of cities miss the subtleties of the power relations between actors working in different offices of knowledge-intensive business service firms, our analysis suggests that the process of constructing power relations in advertising account management work is increasingly rendering powerful offices in cities that do not have Alpha world city status. Because of the demand from clients for advertising services in more and more cities worldwide as part of the trend noted above for marketing management to be decentralized and for clients to emerge in 'new' markets, a broader range of cities has gained strategic importance in transnational advertising work. The discussion of account management work would seem to, therefore, confirm the trends discussed in Chapter 4: the spatialities of advertising work have become more complex over recent years as agencies operate in and through cities in new ways.

Account planning

The geography of the strategic work of account planners might initially be assumed to be similar to that of account managers in that the relationship between the planner and the agency's clients is crucial. Account planners must fully understand a client's aims and, as well as working with account managers to do this, planners need to be able to regularly interact with the client's marketing department. However, there is a complication in relation to the geography of account planning work. Account planners have to straddle the client–consumer boundary and must also have an equally strong 'relationship' with consumers. This 'relationship' with consumers is based both on market research and analysis – quantitative and qualitative market research studies – but, also tacit, encultured and embrained understandings and an empathy with *market-specific* consumers and their attitudes and reflexive behaviours. The following quotation reveals why understandings of market-specific, spatially entangled consumers are so important:

> it gets very complicated very quickly ... when I worked with [motor manufacturer x] in Europe my experience was that the North Europeans and the Southern Europeans would have very different emotional reactions to different pieces of emotional communication. If you were trying to develop emotional resonance, chances were that the Germanic countries reacted very differently to a piece of film or a piece of communication and then the Mediterranean countries. I don't wanna draw broad nationalistic stereotypes but I mean, we're culturally very different. The way I react as an Englishman in American is culturally still very different from the way the Americans react to things.
>
> (Interviewee 27)

However, whilst market-specific understandings of consumers are undoubtedly important, the main advantage of a global agency – the ability to develop integrated communications packages through which the client's brand identity is maintained – cannot be compromised by an approach to account planning that lacks some degree of global alignment. Consequently, the coordination of strategy worldwide is a task that falls to the principal planner in the 'lead office'. The 'lead office' is often the office in which the account manager handling agency–client interactions is based. For the principal planner in this office the challenge is to provide, as one interviewee suggested, 'strategic oversight to make sure that they're all heading overall in the right direction for what downtown Detroit, General Motors, wants but then allowing the decision to be made locally' (Interviewee 18).

Next we focus on the main approaches used by planners working on transnational campaigns to deal with the need for both situated tacit knowledge of market-specific consumers but also agency-wide coordination of a campaign and the implications of this for our understanding of how account planning work operates in and through cities.

Geographies of account planning

The importance of the situated tacit knowledge of consumer behaviours implies that global agencies are increasingly reliant on what might be described as 'islands' of situated knowledge (Amin and Cohendet 1999) in the shape of account planners in different offices worldwide with tacit knowledge of particular markets. Indeed, in order to develop an effective understanding of consumer identities and typologies, and the situated, market-specific influences on emotions, values, attitudes, mores and penchants, account planners prefer to live and work in physical proximity to the consumers they are developing a relationship with. As such, whilst marketing research might form what Asheim *et al.* (2007) call a 'synthetic' knowledge-base that, in theory at least, can be produced by planners working at a distance from the consumer the analysis refers to, and then circulated throughout the firm, account planners also require what Asheim *et al.* (2007) call a 'symbolic' knowledge-base. Such symbolic knowledge is valuable because of the importance of interpretative understandings and sense making in planning work. This 'symbolic' knowledge is much harder to circulate and, in line with the work of Polanyi (1967) on tacit knowledge and more recent work on 'embodied' and 'encultured' knowledge by Blackler *et al.* (1998), could be said to be a form of understanding that requires individual planners to *share* and *participate* in the experiences of the consumers they are trying to understand. Or as Aspers (2010) suggests, account planners need to share the lifeworld of consumers. Co-presence allows planners to learn through everyday practice about consumer experiences and their effect on values, attitudes and emotions.

As a result, co-presence with consumers but also collective learning associated with localization economies, as described in Chapter 3, provide invaluable resources for planners trying to make sense of the spatially entangled consumers in any city. For example, one advertiser interviewed described the importance of account planners developing a situated understanding of consumers when developing a campaign for a car in terms of knowledge of consumer group *and* market-specific roles of cars in everyday life. Agreeing that, in theory, strategic work could be completed using the 'global' model with the analysis associated with account planning and the work of translating this into a strategy for adverts both being completed in one office and then deployed in a worldwide campaign, the advertiser cautioned that, in doing this, 'what people in Chicago or what people in Shanghai would miss is the intensive nature of the automobile, how important it is as a role in people's lives here in Los Angeles as opposed to New York' (Interviewee 29). This quotation highlights, then, the importance of what might be termed cultural relational proximity – understanding of reflexive consumer identities, attitudes etc., developed through the sharing of experiences with consumers and the collective learning facilitated by localization economies. Such proximity is market-specific, needing to be produced for markets that have both inter- and intra-national variations. Consequently, interviewees widely recognized that, 'it would be totally and utterly impossible to create advertising in London or New York for China. It's very hard to create great advertising, powerful, compelling. So you have to be there for that' (Interviewee 5).

The challenge for account planners in the lead office for any campaign is to ensure this need for market-specific understanding of consumers exists alongside degrees of agency-wide coordination that ensure the client receives an integrated campaign, one that develops their brand in a coherent a way as possible. Significantly, this does not mean that it is the job of the principal account planner in the 'lead office' to dictate the strategy of other offices. This would prevent the multiple 'islands of expertise' from responding to situated understandings of markets. Instead, the principal account planner must encourage all planners to work in a collaborative transnational fashion. Such collaboration is often facilitated at the start of a campaign by business travel that allows face-to-face meetings. As one advertiser noted, '[f]irst thing that we did on this project is we rang the people in the key markets in the regions involved and flew them to New York because they're the people who're gonna inform what we do' (Interviewee 7). In addition, and complementing travel, video-conferences before and after face-to-face meetings, email exchanges, the exchange of documents via project intranets and, of course, frequent telephone (conference) calls all allow the collaboration needed in the account planning task (Faulconbridge *et al.* 2009). This collaboration allows planners in different offices to exchange ideas, learn from one another's approach to developing an advert and support each other in the process of developing a strategy for their market. For the principal planner in the lead office this can be a difficult process to manage. But the outcomes can often make the adverts developed in each market far more successful than an approach that relies on each office operating as an isolated island. As one interviewee who had experienced such collaboration suggested:

> Half of my time was spent here [in New York City] sort of briefing the American creatives, and half of my time was over there [in London] trying to put strategic thinking together and building a relationship with the client. So the whole pitch was done in London, we had like four or five meetings. But it was very much a global transatlantic team. What was interesting about that exercise is that we realized how a network can really work to your advantage. The smaller agencies that have maybe four offices worldwide, they're more like hub and spoke models. We have like 80 offices, so we're a truly an international network. What was really good about that is that we chose six of our offices around the world: Spain ... I think it was Madrid, Frankfurt, Milan, Sydney and Mexico. We also asked them to get involved. They weren't part of the core strategic team but they were part of the creative resource. So we had this amazing creative resource happening.
> (Interviewee 2)

The geographical significance of such transnational collaboration should not be underestimated. In contrast to the imperial model whereby advertising strategies were exported worldwide, in particular from the USA and Western Europe, the transnational model involves two-way flows of knowledge and the construction of important relationships that further empower offices outside the Alpha world city hierarchy. In the case of offices in China this is perhaps unsurprising. It is well known in advertising that, despite appearing to have developed 'western' consumer cultures, Chinese consumption practices and habits are unique and even vary between regions within China (see for example, Po 2006). But, it is not just major new consumer 'superpowers' that have become important sites of advertising expertise. Throughout Asia a number of countries previously 'off the advertising map' have emerged as important 'islands of expertise' in global agencies (see Chapter 4 for a list of cities).

Of course, it would be misleading to suggest that all traces of the imperial approach to advertising have been erased. In some scenarios what a number of interviewees described as ethnocentric, US and Western European (and especially London) dominated account planning persists. In addition, as already noted, the continued importance of proximity to the client in account planning work continues to render cities such as New York and Los Angeles and their agglomerations of clients' important sites, and often the lead offices, for significant amounts account management and also planning work. Similarly the dense populations of consumers in these cities and the ability to share their everyday experiences by working in these cities makes the Alpha city hotspots of advertising as important as ever for account planning work. However, the need for situated knowledge of consumers located in markets throughout the world, and the benefits of transnational collaboration, mean that more cities have become strategic nodes in account planning work over the past twenty years. This has not been a zero sum game. The growth of account planning work in, for example, offices in Brazil, China and Thailand has not resulted in work being lost in incumbent offices such as those in New York City, Los Angeles or London. Rather new work has been created as part of the development of the transnational advertising model, the rise of second and third wave advertising and the associated demand for market-specific advertising that is tailored to reflexive, individualistic consumers.

Cities are, then, important for account planning work for a multitude of reasons. City-specific combinations of the presence of clients (market demand) and of consumers give each office in a global agency's network a unique role. We explore examples of different combinations of this client–consumer dyad through later chapters on the cities of New York, Los Angeles and Detroit. Cities are also important because of their labour pools. So far in this chapter little has been said about the need of agencies to recruit talented workers but it should go without saying that the combination of client demand and consumer presence together attracts important pools of skilled labour. This issue is returned to later in the chapter. But it is not just 'territorial' assets that render a city important in advertising globalization. Account planning work is as reliant on the transnational connections forged in agencies and, as such, it is also the connectedness of a city through transnational networks of work that renders its position powerful in a global agency network. This, then, confirms the trends highlighted in Chapter 4 and further reveals just how the advertising production process leads to certain cities having strategic roles in advertising globalization.

Creativity

The work of creatives acts as the linchpin that connects the output of the work of account managers and planners to the final advert that consumers see on the television, in a magazine or on the internet. Consequently, just as account planners must work closely with account managers to understand a client's aims for a campaign, creatives must work closely with account planners to understand the strategy developed for any advert in terms of target audiences, attitudes, values, routines and emotions to be tapped into and, consequently, the 'pressure points' that need to be touched by an advert. Hence, as one advertiser noted:

> I felt that what was necessary was to get the creative, strategic and account groups to really work as one unit ... They live and breathe what's going on together. I recognize that in this day and age you could probably have the creative team sitting out in Los Angeles and still get some good creative while the account group's back here [in Detroit] but in reality I tend to be a believer that the core unit of each of those three need to work together because so much happens every day.
>
> (Interviewee 18)

This quotation reveals the importance of interaction between creatives and other workers in an agency, something that, to a certain extent, necessitates the alignment of the geography of creative work with that of account management and planning work. However, creative work has its own independent drivers that also determine its geography and which deserve further attention here. In particular it is the relationship with *consumers* that is also important for creatives.

For creatives, the principal challenge is to design a format for an advert that captures the target audience's attention in an advertising saturated marketplace, whilst also creating a relationship between the consumer and the product being advertised. This delicate balance has to be maintained in all advertising – the synergism of an advert that both captures consumers' attention and delivers strategically is not just a panacea but a basic prerequisite for a successful campaign – and, therefore, creatives seek to avoid making compromises associated with 'global' adverts. For example, at one level, creatives seek to ensure that the visual media associated with an advert reflects situated norms in the shape of variables such as the dress and actions of people in the adverts, the design and use of products and the background scenery used. As one advertiser described, portraying such situated norms becomes very difficult if an advert is designed for use in multiple markets:

> you've got to black out all the windows so you can't see the driver because then you'll see which side of the road the driver's sitting on. You can't black out the windows in American because it's illegal to have the windshield blacked out. So then you're trying to develop concepts where people aren't driving cars. And then in America you can show a young woman with the car and that's fine but in Croatia that's sexist. So now we're trying to produce commercials where we don't show the buyers of the car because the buyers don't look like the buyers around the world. So where do you stop? Once you've put all those restrictions on it like 'we need to show a car pulling outside a house'.
>
> (Interviewee 27)

The quotation reveals how the 'viscosity' of both products and advertising knowledges places limits on the successful mass exporting of creative work (see for example, Weller 2007). Attending through creative work to subtle market-specific variations relating to the use of products, such as cars in everyday life, their design and specification, is vital in contemporary campaigns. Moreover, the norms of advertising in relation to the balance between information and entertainment in a campaign act as another 'viscous' force that makes the use of market-specific creative work preferable. For example, the factual, content heavy nature of US adverts influences the type of creative work used in campaigns and makes US creative work less effective in Europe where adverts are generally more emotional in their approach. In addition there are also often time–space disjunctures in the consumption of products with, for example, the different timing of the seasons in the northern and southern hemispheres meaning that a campaign developed for the northern hemisphere summer is dated and draws on inappropriate references to world events by the time the southern hemisphere summer arrives (see for example, Weller 2007). Adverts often do not date well in the new reflexive economy and, on many occasions, cannot simply be rerun six months later. Reflecting the arguments developed by Pike (2009) in relation to brands, it is thus important to recognize that it is the inherently situated and spatially and temporally entangled relationships between consumers, products and advertising that define the geography of creative work. But this is not the only challenge creatives face in their work.

The artwork, casting of actors, studio work, post-production editing and other tasks that creatives are responsible for require a network of trusted contractors to be used. These contractors form a vital part of the temporary project team associated with advertising production and, as such, the need to draw on a network of trusted contractors is vital and another key definer of the geography of creatives' work. Indeed, the ever quickening advertising production process described previously has only made the effectiveness of the relationship between creatives and these contractors more important. As one advertiser commented:

> As the ways in which we connect with customers diversify, we don't have all the skills sets in house. If you want to do advert gaming or if you want to

> do product placement or if you want to do something new like that, we don't have the skills sets in house and yet we are definitely having to work with people in the vanguard of those movements ... We are working more and more with those people. Those people are not just executing for us, they're helping us. That is a significant shift.
>
> (Interviewee 7)

Consequently, for creatives, the challenge is to ensure that this need for relationships with contractors, alongside the need to understand consumers in such a way that allows attention-grabbing adverts to be developed for a specific market, is met by the geographical organization of work in global agencies.

Geographies of creative work

As a result of the complex three-way relationship between creatives and account planners, consumers and contractors, the geography of creative work, at first glance, appears to be the most 'local' and city based form of work. Indeed, reflecting existing work on creative industries which emphasizes the importance of cities and the resources they provide both in the shape of markets (access to consumers) and project ecologies (access to contractors) (see for example, Grabher 2002; Rantisi 2002; Scott 2000), for global advertising agencies major world cities act as the key sites of creative production. This can be explained in two main ways.

First, because creatives like planners need to interact with consumers, major cities such as London and New York, as the places numerous consumer groups live, work and play, act as important sites for advertising work. For creatives, presence in such cities allows interaction with consumers on a day-to-day basis as they share and observe the experiences of consumers in the city, something vital for developing effective creative work. Relating to this final point, one advertiser described the importance of cities for creative work as follows:

> So the environment in a creative endeavour where you're stimulated by sights, sounds, things that occur, experiences. It's a more stimulating environment. New York, London, Paris, Tokyo, Shanghai, there's a lot of creativity in these. It doesn't mean that creativity can't exist elsewhere. But if your business is creativity and it's trying to reach a high frequency of turn of creativity, you'll find that in stimulating cities, urban areas where people live closer together and where the opportunity for involuntary stimulation is greater, you'll find that that's a more creative environment.
>
> (Interviewee 1)

Creatives also benefit from localization economies because of the presence of fellow advertisers and the collective sense making described in Chapter 3. Hence 'hanging out' in the city is generally valuable for creatives because of the inspiration it can provide (Grabher 2001).

Second, creative work also renders cities important because of, as already noted, the importance of access to trusted contractors. As a result of the need for 'just in time production' of adverts, coupled with the need to constantly assess the quality of the work of contractors, geographical proximity is preferred between creatives and the contractors they use. Consequently, despite the acknowledged ability of contractors to work at a distance and provide many of the services needed by creatives remotely, interviewees widely agreed that, as one advertiser described,

> There's no reason in the world I couldn't call New Zealand to say 'would you like to bid on the project?' . . . I don't because I have a relationship with a small set of suppliers locally that takes care of me.
>
> (Interviewee 16)

It is important, therefore, to recognize the vital and multifarious role of cities such as London and New York in creative work. However, at the same time it is also important not to assume that creative work is exclusively a city based, 'local', form of work. Leading advertising cities are *not* produced by an entirely 'local' creative process. Creative work also exploits 'global' geographies when appropriate, stretching the creative production process across space so as to tap into resources in other cities and spaces. As such, and like account management and planning work, creative work in global agencies benefits from forms of transnational collaboration and takes place in *and* through cities. For example, creatives benefit from collaboration both with other creatives and with account planners working on the same client account in a different market. For any client account operating in multiple markets, project meetings relating to creative work are held virtually and face-to-face and will involve planners and creatives in multiple markets. Through such meetings creatives learn from the insights of planners and creatives worldwide as ideas are shared and collective learning occurs (Faulconbridge 2006). For example, creatives in New York may learn about consumer reactions to different advertising strategies from creatives in Paris but not as part of a process that involves the transfer of knowledge and strategies from Paris to New York. Thus, there are important differences between knowledge transfer and strategy replication and collective, social and practice based learning (see Chapter 2). The insights gained from such collaborations will be developed by the planners and creatives co-located and working on adverts for the New York market, and used as a springboard for developing market-specific solutions, rather than as a form of best practice or pre-packaged solution.

Relatedly, whilst creatives continue to value sharing experiences and life-worlds (Aspers 2010) with consumers through presence in cities, consumer engagement has also increasingly switched to online spaces during the early part of the twenty-first century. In line with the work discussed in Chapter 3 on user-led innovation, websites such as YouTube and MySpace allow creatives to test run adverts with consumer responses being collected in the form of comments and feedback messages and even suggestions for how to improve a piece of creative work. The consumer really has becomes 'king' in the advertising production process (see for example, Grabher *et al.* 2009) in that early responses and feedback on a campaign posted on a website can determine whether creative work ever makes it to the television screen. As one advertiser described this phenomenon:

> We do a lot of consumer-call creation with ideas, you know, put an idea out there and getting consumers to kind of help take it, shape it, share it, to mix the meaning. I just think you're seeing a real flattening of creativity, a democratizing of creativity. All the models and structures in the big agencies are having to dramatically shift to this new landscape.
>
> (Interviewee 3)

Such user-led innovation involves interactions between creatives and consumers located in virtual cyberspace rather than material space as creatives,

> set up an agency through MySpace and through YouTube and say to consumers 'this is our idea and here's what we want you to do with it, see what you can do with it, see if you can make your own campaign and create it'.
>
> (Interviewee 3)

Cities, therefore, coexist with virtual spaces of user-led innovation as the face-to-face focus group becomes only one means of binding the consumer into the advertising production process. Again this reminds us not to overly privilege the role of the city at the expense of other network spaces of creative work. Indeed, reinforcing this point, whilst contractors located in the same city as the creatives working on an advert offer many benefits, creatives in global agencies also on occasions seek to exploit global talent when it offers advantages over contractors located in proximity. As one New York City based advertiser commented:

> I've shot with a British photographer and he's insisted that he uses his retoucher in London and I sent our art director to London for a week but that's only because that photographer from a quality point of view believes his retoucher is *the best* and therefore I'm buying into that whole process to use that retoucher. Or I've edited in London and worked in London with a director and then used his editor because there are lots of great editors in London. Or I've done it in Los Angeles.
>
> (Interviewee 6)

In sum, whilst complex and multifarious, like the geography of account management and planning work, the geography of creative work is defined by both territoriality and the benefits of operating in particular cities but also network connectivity and the flows between cities and places. We now turn to explaining the implications of such geographies for our understanding of both the operation of global agencies and the role of cities in advertising globalization.

Transnational advertising work: implications for global agencies and cities

The discussion of the work of account managers, account planners and creatives reveals the multiple geographies of the knowledges and practices associated with developing an effective advertising campaign. Reflecting the discussions of cities as both territorial and network formations, the work of account managers, planners and creatives has been shown to be variously tied to the localization and agglomeration assets of particular cities, but also to network connections between cities and global work, as fostered by the transnational organizational form of agencies' in the twenty-first century. Table 5.2 summarizes how these multiple, overlapping and topological geographies influence the work of the three groups of advertising workers.

For global agencies the implications of Table 5.2 are clear: the transnational organizational form and the simultaneous use of the territorial assets of a city and inter-office network collaboration is crucial for the delivery of integrated and aligned advertising campaigns. As such, the ability to use the transnational form to develop campaigns continues to be the main competitive advantage global agencies have over boutique agencies. As one advertiser put it:

> The way I run the business is we're about 'the work, the work, the work', 'creating the most compelling content in the world' and we only have 3 operating principles. (1) secure an unfair share of a limited pool of exceptional talent because it's the exceptional talent that does the exceptional work that breaks through but also that attracts the talent and the clients to a company. (2) Leverage that talent as widely as you can across borders and

Table 5.2 The multiple geographies of advertising work

Job role	*Territorial (city) geographies*	*Network (global) geographies*
Account management	• Face-to-face interactions with clients, account planners and creatives.	• Management via virtual interactions with fellow/lead account managers, account planners and creatives.
Account planning	• Face-to-face interactions with clients, account managers and creatives. • Participation in consumer experiences.	• Virtual interactions with fellow/lead account planners, account managers and creatives allow learning.
Creatives	• Face-to-face interactions with account planners. • 'Hanging out' with consumers. • Regular interaction with sub-contractors in the project ecology.	• Virtual interactions with fellow creatives and account planners allow learning. • Involvement of global experts in project ecologies.

> cultures and brands, forms, words, pictures, sounds, experience and to do that you need some simple processes. (3) Use the network and that's not just about international clients which is half of it. It's also about the fact that there are 16 000 people out there who can help you sort the problem.
>
> (Interviewee 5)

The transnational model does however, as alluded to in Chapter 2, require a rethinking of the power relations between offices so as to ensure that 'imperialism' is replaced by collaboration and cooperation as the strategic role of *all* offices is recognized. This is an ongoing transformation and few working for global agencies would claim that all employees respect one another regardless of their office location. Some account managers, planners and creatives in, for example London and New York, undoubtedly continue to have a superiority complex that leads them to look down on their counterparts in other offices. But the significance of the change occurring should not be underestimated. One advertiser captured the nature of this change as follows:

> I think globalization as I've seen it has worked in two ways. The first way is the negative aspect which is a kind of imperialism that a brand will organize itself from a central hub and oftentimes it's an American hub. The clients are asking us to effectively run their marketing communications almost like a dictatorship which of course has all kinds of horrific ripples that are not good. I can see a second kind of globalization which is very positive, which is the reverse, which is from the street back up. We're doing a lot of projects now where we collaborate with a local market and we get influences and we get more of an alchemy from the local market. The best example I can give you is we recently did something with Thailand. At first, the Thai agency felt it was going to be another case of American imperialism when in fact it was actually the opposite. We were just looking to use some of our tools but really pull all the brain power and all the thinking from that local market. I think the combination of having the right creative tools in terms of getting to ideas and how to execute ideas with the brain power of that local market, that was the right combination.
>
> (Interviewee 20)

This quotation draws out the significance of the transnational organization of advertising work for our understanding of the role of different cities in advertising globalization. Because of the geographies of advertising work described in Table 5.2, more cities are taking on a strategic role in advertising work as defined by their input into the client management, account planning and creative production process. To translate Table 5.2 into an understanding at the level of individual cities, Table 5.3 outlines the territorial and network assets of strategic and successful advertising cities. The most powerful cities – i.e. strategically important and successful in terms of measures of work – are those that are able to exploit all of the territorial and network assets. Exploiting all of the assets is important because the two are not mutually exclusive, but instead feed off one another. Cities only able to exploit some of the resources, for example cities lacking a large client base, are less powerful because of the reduced work generated both in terms of 'local' work (i.e. work for the market the city its located within) and 'non-local' work (i.e. work associated with the control of campaigns being developed by other offices worldwide for a 'local' client).

In terms of the debates about the role of territorial and network assets, the implications of Table 5.3 are that, as the world cities discourse claims, the most strategically important sites of work in the twenty-first century are indeed those cities connected into global networks of trade and knowledge flows. A city with all of the territorial assets listed in Table 5.3 may well have a solid advertising industry, but without the network connections also described in the table, the potential to grow advertising employment, and the resilience to economic change of the advertising industry in the city, are likely to be limited because of a reliance on 'local' clients and markets.

The reading outlined in Table 5.3 of the determinants of a city's powerfulness in global advertising geographies is, therefore, helpful for analysing the global map of advertising work in the twenty-first century outlined in Chapter 4. It suggests that the combination of the emergence of indigenous clients and the need for market-specific advertising and the resultant elevation of the value of the territorial resources of cities once off the global advertising map (i.e. the importance of the skilled workers' knowledge about market-specific consumers in cities such as Bangkok) alongside network flows of trade (i.e. client demand for multiple adverts produced to suit geographically heterogeneous markets in places such as Thailand) have begun to lead to new distributions and divisions of advertising labour and work.

Conclusions

In this chapter we have developed the conceptual framework outlined in Part I of the book through a detailed empirical analysis of the practices and geographies of advertising work. By focusing on the multiple geographies of account man-

Table 5.3 The territorial and network assets of successful advertising cities

Territorial assets	*Network assets*
• Strong client base (creating local and non-local work) • Deep pools of skilled labour (the ability to develop effective market-specific advertising) • Large consumer markets (creating demand for market-specific advertising and available for 'study' by account planners and creatives)	• Leadership in global campaigns (generating advertising work that is non-local in remit) • Membership of global campaign teams (flows generating work associated with developing advertising for the 'local' market)

agement, account planning and creative work we have begun to reveal the role of cities in advertising globalization, not as exclusively territorial formations but as sites that advertising work occurs in *and* through. Based on this reading, in the final section of the chapter we began to speculate on the ways that different cities pin down advertising globalization depending on their territorial and network assets and suggested that this helps explain the changing map of global advertising work outlined in Chapter 4. Of course, in doing this we have raised as many questions as we have provided answers. For example, we have not yet considered how the changing power relations we have described have affected incumbent cities – places like London and New York City – that have historically been the 'command and control points' (Sassen 2000) of advertising globalization.

In Part III of the book, drawing on original qualitative research findings, we, therefore, consider how these dynamics have affected three incumbent US cities – New York, Los Angeles and Detroit – as their strategic role in advertising globalization changes in line with the evolution of the transnational agency model.

Part III

Agency-city relationships in advertising globalization

6 New York City

From centre of global advertising to a global advertising centre

> New York City has always been home to artists, musicians, writers, and actors. The city has always been a hotbed of creative and intellectual breakthroughs, driven by creativity and talent. Although different industries have powered New York and have taken on great importance throughout history, creativity has always been a part of New York City's raison d'être.
>
> Currid (2006: 333)

New York City, since the invention of mass consumerism in the USA, has been synonymous with advertising, 'Madison Avenue' being the home of many leading global agencies from the 1920s until the 1980s (Leslie 1997a). But, the role of New York City in advertising globalization is not simply one of first-mover advantage and the continued hegemony of the city. Rather the story is one of flux and evolution. After initially being *the centre of global advertising*, the 'imperial command and control centre' when campaigns were designed in New York and ran worldwide, from the early twenty-first century the city has become *one amongst many global advertising centres*, which collaborate and cooperate to deliver multi-market campaigns.

In this chapter, we analyse the evolving role of New York City and use the insights gained from the research to further develop our conceptual argument about the transnational organization of advertising agencies and the role of territorial and network assets in determining the importance of a city in the global geographies of advertising work. Drawing on ideas about the role of transnational collaboration and embeddedness, agglomeration, localization, project ecologies and the evolving geographies of the advertising production process outlined in Chapters 2, 3 and 5, we argue that New York City acts as an exemplar of a world city that has seen its role reconfigured as a result of the development of the transnational agency form. But, through the optimum exploitation of both territorial and network assets, New York City has, we contend, manufactured itself a continued and leading role in advertising globalization, albeit a different role in the twenty-first century from the role played for much of the twentieth century.

New York: the iconic centre of global advertising

The history of New York City as *the* centre of global advertising conforms in many ways to the 'imperial' model discussed in Chapter 2. Six of the top ten global agencies were born in New York City during the twentieth century (see Table 2.2). This was primarily a result of the city's role as a centre of manufacturing which meant that the advertising industry benefited from the wide range of local clients that were seeking agencies capable of advertising consumer goods worldwide using a single 'global' advert. New York City also benefited from a wide range of clients because of the television networks based there. Together these two influences turned New York agencies into manufacturers' partners in 'Americanization' and the exporting of consumer society (Leslie 1995; Perry 1990).

As such, New York City was the 'place to be' in the twentieth century for advertising agencies, their clients, their networks of externalized suppliers and for the individuals seeking a career in advertising. As one advertiser interviewed noted about the heyday of New York City as *the* centre of global advertising:

> When the industry developed in the US, there were good reasons why it made sense to be in New York. There were lots of clients here, a lot of the media ownership was here, so given the importance at that stage of being able to buy distribution, this was a good place to be and there was plenty of talent, lots of people.
>
> (Interviewee 5)

The quantitative analysis presented in Chapter 4 shows that New York City has built on this historical role to maintain its position as a major centre of global advertising in the early twenty-first century. Table 6.1 summarizes a number of

Table 6.1 Summary statistics relating to New York City's advertising industry in the twenty-first century

Billings 2001 (US$ millions)	57,237.60
Global rank by billings (2001)	1
Lead over number two ranked city by billings	48%
Number of headquarters of top six holding companies/value of companies' income (2008)	2/US$20.32 billion
Closest rival in terms of headquarters of top six holding companies (2009)	London (1 headquarters, income US$13.60 billion)
Number of top six holding groups with one or more agencies in New York	6
Number of advertising agencies (NAICS 54181) 2001/08	1,084/986
Total employment in advertising agencies (NAICS 54181) 2001/08	33,542/34,237

Source: analyses presented in Chapter 4 and US Bureau of Labor Statistics Metropolitan Area Occupational Employment and Wage Estimates New York, NY PMSA key statistics regarding agencies, billings and employment in New York City in order to emphasize its continued significance in the twenty-first century. At first glance, such data suggests that New York City continues to be what Sassen (2006a) would call a 'command and control centre' for global advertising: a leading world city with locally based advertisers in the city coordinating advertising work worldwide. However, such statistics and descriptions hide a more complicated story about the way New York City's role in global advertising work has evolved over the past twenty-five years or so.

Agglomeration, localization and the success of New York City in the 2000s

Since Leslie (1997a) wrote about the abandoning of Madison Avenue, many advertising agencies have returned to Madison Avenue with Midtown Manhattan now being the main location of global agencies (see Figure 6.1). Of course, if non-global agencies were added to this map the geography of the industry would look very different. According to the US Bureau of Labor Statistics, in 2008 there were 34,237 workers employed in advertising agencies in New York City. But 86 per cent of agencies in New York City employ fewer than twenty people. Global agencies, according to our research employ hundreds, and in some cases thousands, of advertisers in their New York City offices, but are exceptional, being part of the 1.2 per cent of agencies that employ more than 100 individuals. As a result, a map of all agencies in New York City would reveal advertising workers are actually spread across all parts of Manhattan Island in small boutique agencies.

The first, and most common, explanation of New York City's persistently dominant role in advertising globalization in the twenty-first century relates to the agglomeration advantages of the city in the shape of the continued prevalence of demand for advertising services because of an abundance of clients (see for example, Leslie 1995, 1997a). In some cases this is associated with a form of path dependency. Many clients located in New York City at the time of the birth of global advertising, or operating through the city and using the services of agencies there to develop 'global' campaigns during the height of Fordist consumerism, have maintained exclusive relationships with an agency originating in the city. For example, BBDO New York has managed the PepsiCo brand since the start of the 1960s. Indeed, the fact that agencies in New York City manage an exceptionally wide spectrum of consumer brands, from food and beverages to cosmetics, household goods and services such as banking, insurance and tourism, is a result both of historical relationships and the fact that the headquarters, or marketing headquarters at least, of so many US consumer brands continue to be located in, or close to, New York City. As one interviewee summarized the advantage this situation gives the city:

> New York has always been a competitive market. New York is not a one product category advertising community. I suppose when Boeing has problems, Seattle suffers. When automobiles have problems, all the automotive

suppliers suffer. And I don't think that's true in markets like New York, or Chicago or Los Angeles, because they're so diversified.

(Interviewee 16)

Key clients managed by the New York City offices of global agencies we studied included Bacardi, Colgate, Excenture, Campbells, Xerox, Proctor & Gamble, Bank of America, Gillette, Unilever, Johnson & Johnson.

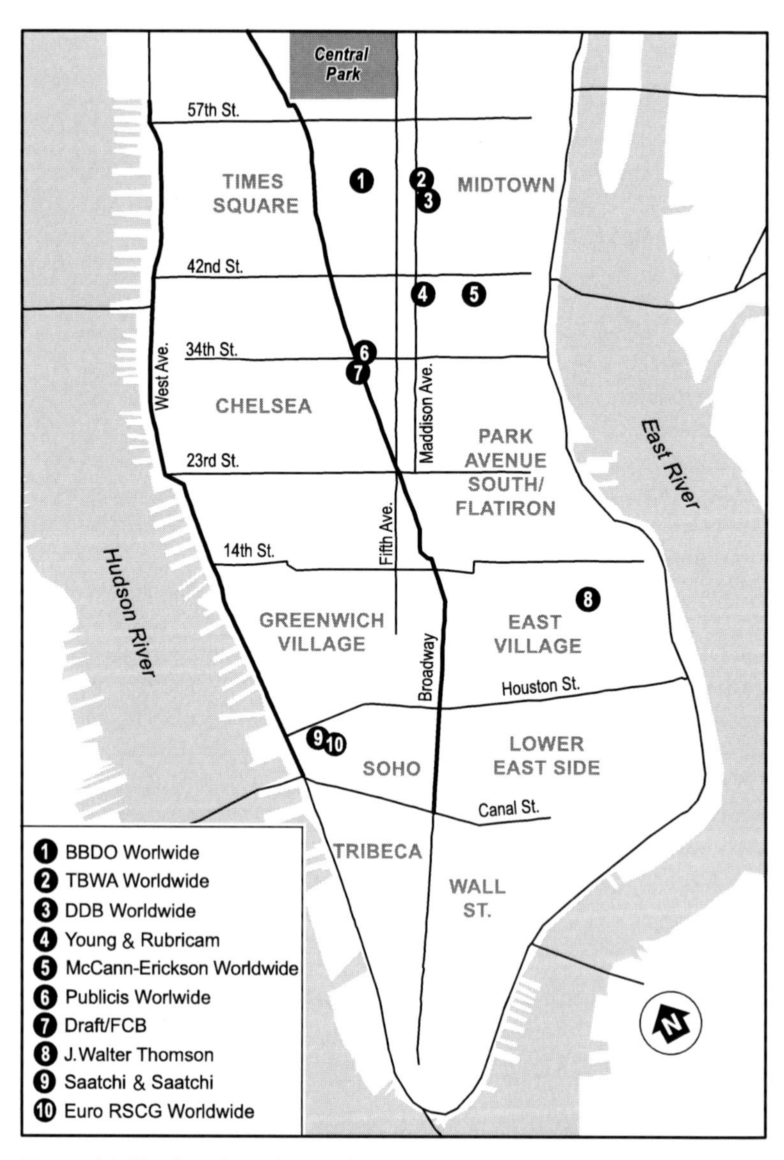

Figure 6.1 The location of key global agencies on Manhattan Island, New York City (source: authors' research).

In addition, New York City's role in global advertising is a result of the strength of the city's localization economy as a territorial asset that helps generate the knowledge and expertise needed to design effective adverts. Indeed, Currid (2006) argues that such creative localization economies are actually more important than agglomeration economies in making New York a key world city. At one level the city houses an advertising milieu, or 'project ecology' (see Chapter 3), made up of both freelance workers and companies with 'related variety' (Frenken *et al.* 2007), such as film and music studios. Table 6.2 shows the high levels of employment in some of the key industries that advertisers draw on in the production process. The strength and depth of the project ecology in New York City, comprising many individuals and firms that provide crucial services that are part of the advertising production process, both facilitates repeat relationships and the development of trust and mutual understanding, and effective searches for talented workers (see for example Ekinsmyth 2002; Grabher 2002).

The existence of such a strong and deep project ecology results both from the demand for the work of such individuals and firms, but also from the iconic status of New York as a city. As a city reputed both within the USA and worldwide as the 'place to be' for aspiring advertisers, but also creative workers more generally, New York City draws in talented workers in ways that mirror Florida's (2002) somewhat contested 'creative class' thesis. Our aim here is not to engage explicitly in debates about regional development and the role of urban regeneration and resources, such as theatre and art, in attracting talented workers. We do, however, return to such issues later in this chapter and in the conclusion to the book to explore the implications of our understanding of what makes a successful advertising city for such approaches. For now we simply note that there is little doubt that the pools of talented workers located in New York City are one ingredient in the recipe that continues to make New York City a centre of advertising work. As one advertiser commented:

> The importance about being here – it should be our greatest advantage – is just the sheer number of creative people. Just the fact that this is a centre for

Table 6.2 Employment in 2008 in New York City in key industries making up the advertising project ecology

Actors (NAICS 272011)	4,210
Graphic design (NAICS 271024)	21,870
Photography (NAICS 274021)	3,460
Film production (NAICS 274031)	1,790
Musicians (NAICS 272042)	6,720
Set designers (NAICS 271027)	920
Post-production editing (photography, music and film) (NAICS 274032)	2,290
Lawyers (NAICS 231011)	70,660

Source: US Bureau of Labor Statistics May 2008 Metropolitan and Nonmetropolitan Area Occupational Employment and Wage Estimates, New York County.

> world creativity and world entrepreneurialism and all that. So New York and London definitely have those advantages. I lived in Minneapolis for ten years. It's very hard to get a creative person to move to a place where it's 20 degrees below zero.
>
> (Interviewee 3)

At another level, New York City also has another major localization advantage in the form of the large and broad consumer markets present in the city that producers of consumer goods wish to target via advertising. Reflecting the ideas of Mattelart (1991) that there is more diversity between Midtown Manhattan and the Bronx than there is between Midtown Manhattan and the seventh Arrondissement in Paris (see Chapter 3), the fact that advertisers are able to *share* and *participate* in the experiences and 'lifeworlds' (Aspers 2010) of the diverse array of consumers that they are trying to understand, and target advertising at, is a major benefit of locating in New York City. As such the city has a territorial asset in the shape of access to a wide range of consumer groups which can be learned about and then targeted in campaigns. Advertisers described how by both 'hanging out' in the city, in cafes, bars or shopping malls, and by attending major public events, such outdoor live music or television shows, it is possible to capture important insights into a range of consumer audiences. As one advertiser described the benefits of such learning and knowledge generation:

> you're stimulated by sights, sounds, things that occur, experiences. It's a more stimulating environment. New York, London, Paris, Tokyo, Shanghai, there's a lot of creativity in these. It doesn't mean that creativity can't exist elsewhere. But if your business is creativity and it's trying to reach a high frequency of turn of creativity, you'll find that in stimulating cities, urban areas where people live closer together and where the opportunity for involuntary stimulation is greater.
>
> (Interviewee 1)

One agency we studied even went as far as organizing its own event in Times Square for one of its clients, a major US retail chain, both to promote their products, but also to gauge different consumers' reactions to particular products and types of promotion. As an interviewee from the agency described:

> for their thanksgiving sale campaign, we had [celebrity x]. The idea was if he can last the week-end, you can take a hundred kids shopping for free. We had him in a gyroscope spinning in Times Square. It was a live event that was also filmed and that you could watch online and enter the competition.
>
> (Interviewee 5)

In sum, the combinations of agglomeration and localization advantages render New York a city with national and international competitiveness in terms of territorial assets. Both the demand for advertising because of agglomeration advantages and the ability to deliver effective and innovative campaigns because of localization advantages, mean that agencies are able to generate the impressive level of billings reported in Table 6.1, and command a leading role in the US and global geographies of advertising work. However, a caveat does need to be added to this argument. New York City's role and advantages have come under threat in recent years because of competition in global advertising networks of relations.

Twenty-first century challenges and the new city of networks

New York City's network connectivity is a crucial part of the explanation of the strength of the city from the mid twentieth century Fordist era of 'global' advertising onwards. Table 6.3 captures some of the key indicators of New York City's contemporary connectedness in global networks of advertising. In the mid twentieth century, agencies used network connectivity to export adverts as part of 'global' strategies. Offices in other countries primarily used connectivity with New York City as a means to receive, adapt and deploy 'global' campaigns. But as the analysis in Chapter 4 suggests, this role of the network connectivity of New York City has been changing and, whilst the city continues to be ranked in first place in terms of global connectivity, the relative strength of the city's connectivity has slightly declined in the early twenty-first century. One change relates to the city's domestic role as a centre of US advertising, another to the city's role as *the* centre of global advertising. We deal with the domestic challenge first.

In many ways the USA is exceptional in terms of the geographic arrangement of advertising work. The existence of multiple centres of advertising is something seen in few other countries worldwide, China being one of the other exceptions (see for example, Po 2006). During the twentieth century cities such as Chicago, Detroit and Los Angeles developed advertising industries to serve local manufacturers of consumer goods and their US advertising needs. To a certain extent such a driver of a multi-centred advertising system still exists in the

Table 6.3 The connectivity of New York City in terms of advertising work

Ranking of connectivity via office networks of global agencies	1.0 (nearest rival London ranks 0.74)
Score of importance of the city in terms of connectivities associated with European advertising work	2.23 (0.45 behind the leading city London)
Score of importance of the city in terms of connectivities associated with Asian advertising work	1.10 (3.19 behind the leading city Tokyo)
Number of headquarters of ten largest global agencies	9

Source: analyses presented in Chapter 4.

twenty-first century. As one interviewee noted, 'the largest clients, they tend to be located here in New York City. But, you could probably do the same analysis of how many of potential clients have offices outside of New York' (Interviewee 3). More recently, however, the multi-centred system has also taken on a new role.

The multiple US offices of global agencies, and the multiple offices of major US domestic agencies such as Doner and the Richards Group, have a dual role in the contemporary post-Fordist era. First, they continue to provide services to 'local' clients seeking to advertise consumer goods across the USA and/or the world. Second, they also give agencies the ability to develop advertising tailored to the geographical heterogeneity and place-specific entanglements of consumers within the USA. Whilst at one stage in the twentieth century the idea of New York based agencies serving all of the needs of their clients throughout the USA (and the world) might have seemed feasible, in the twenty-first century the intra-US role of the New York offices of agencies is very different. These offices act as both 'soft dictators' (see Chapter 5) managing pan-USA campaigns, and as the collaborators on pan-USA campaigns controlled by other offices in cities such as Chicago and Los Angeles. This role has emerged because of, as one interviewee summarized using the example of the market for cars in the USA, the need for advertising tailored to the diverse consumer cultures across the USA:

> New York, Massachusetts, Florida, places like that: mainly European cars again. California, Washington D.C., like around the Seattle area … you go inland here, even here you go to a place like Bakersfield which is mainly farming you'll see a lot of American cars. So the more urban, well travelled cosmopolitan people, they don't wanna be seen dead in an American car. And then you go to places even within California, east of the cascades or Washington or something, and all you'll see is American cars. Very, very strong demarcation.
>
> (Interviewee 30)

The geographical variability in consumer preferences described and, implicitly, the different 'entanglements' (Pike 2009) affecting consumers' relationships with cars alluded to in this quotation, mean that the agglomeration and localization advantages of New York do not necessarily afford the city a command and control role in US advertising work. It is not always effective to design campaigns in New York City and simply export them across the USA. Instead, collaboration with situated offices in different regions of the country is needed to ensure, first, that campaigns managed by the New York City office work effectively across the USA and, second, that advertising appropriate for east coast consumers can be developed for campaigns managed, for example, by the Chicago or Los Angeles offices. Whilst new forms of user-led innovation mean it is not always necessary to 'be there' to engage with and understand consumers, at present it seems that in advertising the benefits of 'being there' and sharing the lifeworld of the consumer continue to render using an *in situ* office invaluable in developing effective advertising suited to the needs of complex, geographically heterogeneous and entangled consumers in different markets. This is the first important dimension of the new *cooperative and collaborative* relationship with other cities, in the USA and worldwide, that New York City relies on in the twenty-first century for its success, something different from the imperial command and control relationship the city had with other cities in the mid to late twentieth century.

The second dimension of New York City's collaborative and cooperative relationship with other cities relates to the decline of the city's role as an imperial exporter of advertising to the world. As Chapter 2 outlined, the early stages of advertising globalization were led from the USA and New York City in particular. This involved an imperial export model whereby 'global' adverts were developed in the USA and deployed by post box offices. This configuration rendered New York City *the* centre of advertising globalization. However, the latter part of the twentieth and early twenty-first century saw this role challenged as part of the switch to the post-Fordist, 'transnational' models of advertising globalization. As the need for advertising that responds to reflexive, individualized and spatially heterogeneous and entangled consumers grew, global agencies had to respond and reconfigure the way they managed worldwide campaigns. As a result, the New York offices of global agencies had to take on a new role as part of a more collaborative and cooperative agency network. The latter part of the twentieth and early twenty-first century, therefore, saw the New York offices develop a new structural role in global agencies' networks. As one creative director put it, a realization occurred that:

> There needs to be complete refocus on what agencies do. Agencies grew up in the old post-war American economic imperialism where large American companies expanded across the world. I was part of an agency at the time that basically expanded where its clients wanted it to go. So there are legacy issues to do with regional structures of agencies that have not been fully worked through.
>
> (Interviewee 7)

Working through these legacy issues involved repositioning the New York offices of global agencies as the 'soft dictator' for US clients' transnational campaigns (see Table 2.4 on the differences between global, 'imperial export' and transnational, 'collaborative and cooperative'). It also involved developing a new role for the New York offices, as collaborators on campaigns led by offices throughout the world. This latter role involves developing adverts for markets on the east coast of the USA for Asian, European, Indian and South American clients. Toyota was one of the first clients to require such cooperation and collaboration from the New York offices of global agencies and the number of non-US and European clients has multiplied in the early part of the twenty-first century. As one advertiser described this phenomenon:

> As Asia and India become far more important for American clients then the perspective from New York shifts. We find ourselves – and I think it's great – working far more with colleagues in different markets, different offices and exchanging ideas and information far more than we used to do as a lot of those markets emerge as a force in their own right.
>
> (Interviewee 4)

In addition, the New York offices of global agencies also rely on collaboration and cooperation with other offices to enhance the effectiveness of campaigns developed. Whilst the localization economy of New York City provides many assets in terms of developing effective campaigns, there is also an increasing reliance on flows of expertise from throughout the world into the city. This can involve, first, virtual flows between offices, the nature of which was captured by one account planner as follows:

> In the UK, [drink x] is very much a lady's drink. You would never see anyone drinking [drink x] in the US. Usually the advertising is based in my office. We've engaged the London office so both creatives in London and New York worked together and we tested in both markets. [Drink y] is a drink that is quite popular in Asia and in Australia and so I used Sydney to help on that.
>
> (Interviewee 6)

Such collaboration and cooperation is vital, both for dealing with spatially heterogeneous consumer behaviours and entanglements when running a global campaign but also for gaining inspiration for campaigns being developed in New York City for the east coast USA market. Learning from other advertisers worldwide allows planners and creatives in New York City to develop innovative and effective campaigns that draw on both 'local' expertise in the city's localization economy but also the expertise of advertisers in other cities worldwide (see for example, Faulconbridge 2006). But whilst such virtual collaboration and cooperation is valuable, taking place using email, videoconferencing and telephone calls, or involving occasional business travel as an individual is flown in to help on a project (see for example, Beaverstock *et al.* 2009), it is not only these inward flows of expertise that maintain New York City's pre-eminence in advertising globalization.

One of the strengths of New York City's localization economy is the depth of the talent that global agencies can draw on in the labour market. At one level this is talent drawn from across the USA as individuals migrate to the city either because of its attractiveness to creative workers or as part of a deliberate career strategy and search for work in advertising. However, at another level the depth of New York City's talent pools in the twenty-first century is a result of the city's global connectedness. In all of the agencies studied in New York City a significant proportion of the workforce was non-US born. This might be expected in a cosmopolitan city such as New York in which one-third of the population is non-US born and over 170 languages are spoken. Indeed, such diversity has been identified by Sassen (2000, 2006a) as one of the defining features of a world city and New York City in particular. Foreign born workers in advertising fill key positions in account management, planning and creative departments and in the latter two departments in particular. The presence of these non-US born workers was viewed by all of the advertisers interviewed as essential for the success of New York City based agencies in the twenty-first century. Most notably these workers were seen as bringing innovative advertising ideas for products such as cars and beverages that US agencies had been advertising for many years using the same, increasingly stale, formats. As one advertiser notes:

> We're running out of good ideas here so quickly because we've come up with so many. It's gonna be very hard to come up just from drawing on our resources with a good new idea. If you inject French art direction, or French music, or English humour, Italian style whatever ... and even Japanese sense of design, very clean, all of sudden your minds, your opportunities are opening up incredibly. As long as we keep hiring and thinking of etc. etc. though an American lens, you're basically under-nourishing the patient ... Actually it's an intellectual lie, we're lying to ourselves when we say 'we're being as creative as we can be' when we're only speaking to Americans. It's stupid. Absolutely stupid.
>
> (Interviewee 30)

The attraction of New York City to overseas workers, in the same way that the city is attractive to US workers, helps ensure a steady flow of foreign talent into agencies. But agencies also supplement such flows by deliberately searching for talent in the hotspots of advertising worldwide. All of the agencies studied deliberately used their network of overseas offices as a tool to search for the best talent and then bring such talent to New York City. As such, the fact that most offices in global agencies' networks are more than just post boxes and, therefore, employ a diverse array of talent both to develop campaigns for US clients, and to lead worldwide campaigns for their 'local' clients, provides an opportunity to use collaborative and cooperative network ties to enhance the ability of the New York City office to develop world class campaigns. As one account planner noted:

> Just looking at some of products from [agency x] in terms of creativity, two of our best agencies in the world are in, first, Sao Paolo. [Agency x] in Sao Paolo, if you look at it in terms of creative awards, they're an extraordinary agency. And then the second place that comes to mind particularly is Bangkok which is an extraordinary agency too. It's extraordinary creative talent, great.
>
> (Interviewee 4)

As a result, one agency we studied had schemes which involved both US workers spending time in overseas offices in places such as China and Thailand, to help develop the capabilities of advertisers in the overseas office *and* learn from their approach to campaign development, and advertisers from other offices spending a period of time in New York. As one interviewee suggested:

> I'm hoping that when we send [advertiser x] to Shanghai he's gonna take the practices, the methods, ways for working and expertise that we've developed out to that market and adapt them to the local culture and hopefully make them better. I also want him to come back. In two years time or whatever it is it'll be fantastic to have him come back. The planning director in Shanghai is coming here to spend a month in the department working with my planner. I had a planning director from Moscow here for six weeks earlier this year. We have two creative teams from Brazil who are now in New York. I have a group of planning directors from Mumbai here who have been here for two years. I'm talking to two people in London about coming here. So we are as an industry getting much better at moving talent around.
>
> (Interviewee 4)

Such examples of the movement of advertisers within an agency's network were common and the quotation above is typical rather than exceptional and highlights the way flows of labour connect the New York City offices of global agencies into wider relational office networks. The benefits described in the quotation above in terms of learning fit what Bartlett and Ghoshal (1998) describe as a 'transnational' model of organizational learning and facilitate what in Chapter 2 was portrayed as a transnational model of advertising. Such an approach to advertising cannot be described using a language that emphasizes geographical boundedness and the role of an office/city acting in isolation. The success of the New York City office in generating billings through innovative and effective advertising is now as much dependent on its relationships with other cities and 'global work' (Jones 2008) as it is on the territorial assets of the city.

As such, by connecting into wider relational networks it is possible for New York City to maintain a competitive advantage over other cities and further enhance the advantages accrued from the agglomeration and localization economies of the city. This means that sustaining the number one position in rankings of advertising billings generated is a different type of achievement in the early twenty-first century than it was in the mid-twentieth century. Specifically it is an achievement born out of the exploitation of the city's network assets to generate exports of advertising and demand for advertising, and the capability of the New York office to deliver effective advertising.

Implications

The discussion of the role of New York City and the agencies based there in global advertising work in the twenty-first century has a number of implications. First, it provides important insights into the way global agencies are structured and the role of the transnational agency form in delivering worldwide advertising campaigns. New York City was historically the home of global advertising, both because of the emergence in the city of a disproportionate number of global agencies and because of the role of US producers of consumer goods in generating demand for worldwide campaigns. As a result, relationships between the New York City office and other offices worldwide were defined by the archetypal 'command and control' power relations. With the exception of a few influential offices in key European markets – London and Paris in particular – the New York offices of global agencies understood their role to be one of exporting advertising to the world as part of an Americanization process tied up with Fordist consumerism.

The snapshot provided here of the work completed by the offices of global agencies based in New York City in the twenty-first century reveals a changed role in advertising spatial divisions of labour. In particular, the empirical material analysed points to the role of New York City based offices in exploiting the territorial assets of the city – the agglomeration of clients, the large and broad consumer markets and the localization economy facilitating innovation through the city's project ecology – alongside and in tandem with the network assets of the agency and city – global work and the inwards and outwards flows of people, knowledge and trade. It is the exploitation of both these territorial *and* network assets that defines the work of the New York offices of global agencies, in particular in terms of both exploitation of the inward flows of trade (i.e. work in the shape of demand for campaigns for the east coast of the USA from producers of consumer goods located inside and outside the USA), and of knowledge and expertise (people and ideas and inspiration). We argue that the wider role of the New York office in global agency networks in the twenty-first century is, therefore, now defined by *collaboration and cooperation* with other offices.

Most importantly, the development of such a transnational model has changed New York City's role in the twenty-first century 'world order' of global advertising. The switch to a transnational model has forced advertisers in New York City to revaluate their role and power in global advertising. This reflects the transition of New York City from *the* centre of global advertising to *a* centre, amongst many, of global advertising. Offices that were once post boxes for campaigns developed in New York City have now developed a number of strategic resources. These resources exist, first, in the form of 'local' clients that generate intra-agency trade when they seek worldwide campaigns. Second, the resources exist in the form of talent and expertise relating to the geographically specific behaviours and entanglements of consumers that are vital for developing situated, place-specific advertising campaigns. Because offices in global agencies' networks, including the New York City offices, need to access these two resources, places such as Sao Paulo and Mumbai are rendered powerful.

This does not, however, necessarily mean the New York offices of global agencies have undergone a process of disempowerment. The city is still incredibly powerful in the process of advertising globalization (see Table 6.1). The growing power of offices in cities such as Sao Paulo and Mumbai has, then, led to qualitative changes in the power relations between the New York offices of global agencies and offices located throughout the rest of the world. These changes in power relations relate to the development of *new* resources that are available to other cities, for example in the form of 'local' clients who operate worldwide, and which help to empower cities such as Sao Paulo and Mumbai and reconfigure the power geometries of advertising globalization generally. As such, offices like those in Sao Paulo or Mumbai have not disempowered the New York City offices of global agencies by 'stealing' resources from New York. As Allen (2003) reminds us, power relations are not a zero sum game. The growing power and influence of cities such as Sao Paulo and Mumbai should be seen as part of a process of other cities reconstructing their global network relations through the development of new resources that qualitatively change relationships between the New York City offices of global agencies and the world. As such the power of the New York City offices of global agencies, once associated with an imperial strategy, is reconfigured as the resources of offices in other cities lead to the construction of a new form of power geometry, defined by being a collaborator and co-operator on transnational campaigns run out of cities worldwide.

As a result the New York City offices of global agencies remain powerful, but because of the combination of what network assets (relational flows into and out of the city) *and* territorial assets (agglomeration and localization advantages) do for the city. Or to describe such a phenomenon in terms of the nature of advertising work, this means that, indeed, power is defined by the way the New York offices act as leaders of transnational campaigns for US clients. As one advertiser noted:

> We have [US client x] which is a big consultancy business. That's global, that's run out of New York. And we have [US client y]. [US client z] is the other one. [US client z] is run globally out of New York. Everything. If you go down to the 4th floor, you'll see an organization that's almost an agency within an agency. It truly is a powerhouse. It's a big big client. That's probably our biggest client on the world stage, we're in every country.
>
> (Interviewee 2)

But, at the same time, New York offices also become powerful because of their vital role in collaborating and cooperating to serve non-US clients' needs for successful advertising on the east coast of the USA. As one interviewee put it:

> We have [non-US client x] in this office it's really run out of London globally and we help do work for the U.S. market. That's a case where we'll show work to the global creative director in London and get his approval to run it because he's really the shepherd of what the voice of the brand is. That's a case where he listens to us; that's a case of that soft dictatorship where he listens to the New York guys and goes 'okay if that seems appropriate for New York, then let's do that as long as it's on the brand's voice'.
>
> (Interviewee 6)

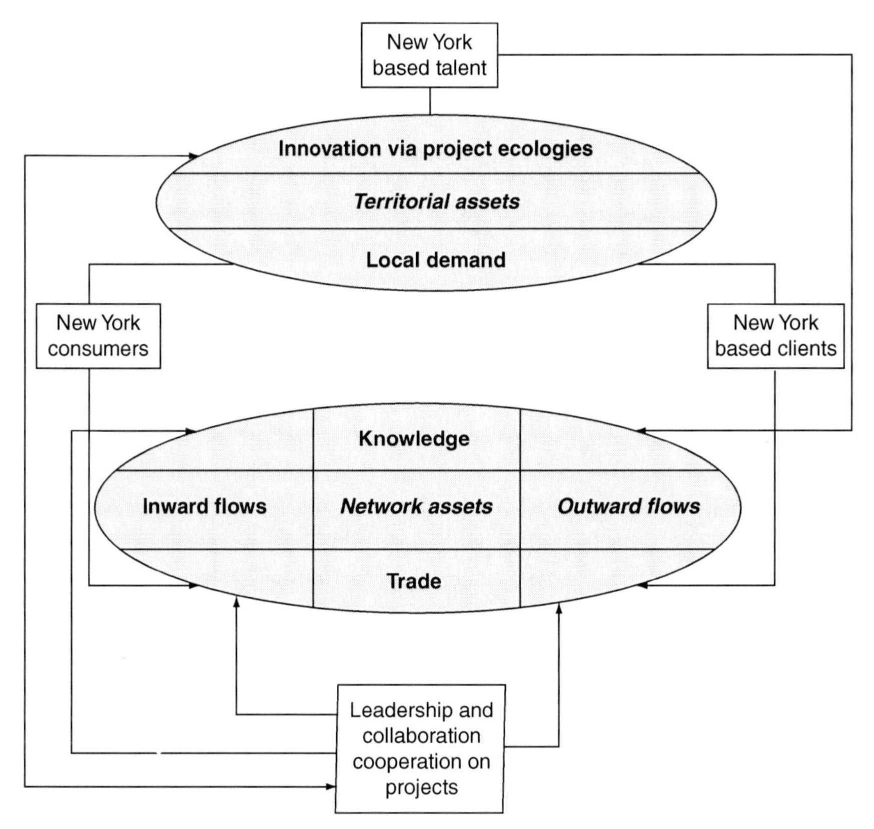

Figure 6.2 The territorial and network assets of New York City and the interrelated and inseparable ways that make the city the leading global advertising centre by billings.

Returning to the discussion of the role of territorial and network assets in defining the power of a city in advertising globalization, New York City appears to fit, then, the model in Table 5.3 of the powerful world city. Only because both the city's territorial and network assets are so strong is New York the leading global advertising centre, as judged by billings. As such, this confirms the suggestions of work on global and world cities that highlights the importance of networks and flow in defining the contemporary geography of service activities (see for example, Castells 2000; Sassen 2006a; Taylor 2004). But, in doing so it begins to also move towards further specifying the nature of such flows, how they are generated and how they interact with the territorial assets of the city. In particular the discussion in this chapter begins to tease out the ways that territorial and network assets are intimately interrelated and in many ways inseparable. Figure 6.2 demonstrates how this inseparability renders New York City powerful. It shows how the city's world leading billings are only explicable with reference to:

1 the way territorial assets generate advertising work (because of demand from clients in the city for advertising and the ability of the project ecology to produce effective advertising);
2 the way territorial assets also produce network assets (the existence of a local client base means the city has control of worldwide campaigns and the city's innovative project ecology means advertisers in New York are called on for input, via flows of trade, knowledge and people, into campaigns developed and managed elsewhere);
3 the way network assets (re)produce elements of the city's territorial assets (flows of trade, knowledge and talented workers due to collaboration on projects led by other offices reproduce the city's project ecology).

Conclusions

New York has evolved from being *the* global centre of advertising to being *a* centre of advertising globalization. This change has involved both a change in the nature of the work completed by the New York offices of global agencies and, at the same time, has led to reconstructed power relations as imperial, global command and control models of advertising work are replaced by transnational collaborate and cooperate models. New York City prospers because of its complex business ecosystem that survives and reproduces itself within the city *and* through collaborative and cooperative connections with other cities.

As such, at one level the idea that New York City is a creative city, able to attract a creative class which in turn ensures economic prosperity holds water. Flows of workers into the city, from the rest of the USA and the world, are indeed one ingredient in the city's success. But, in the treatise of Florida (2002) such flows are independently assumed to generate trade as firms follow workers. Whilst this is partially true in the case of New York City – the presence of talented workers in the city's project ecology attracts clients seeking advertising that will effectively engage east coast USA consumers – it is only because the attraction of the city's workers is coupled to pre-existing demand in the shape of a deep pool of clients and, in the case of advertising, a broad and deep pool of consumers that a successful, world-leading advertising centre is produced. Remove any one of these contributing factors and the interactions between them and New York City's strength as a centre of advertising is reduced. Reflecting the suggestion of Pratt (2008) that there must be a pre-existing market for the creative class to serve for a city to be successful and that attracting talented workers alone is not enough to ensure economic prosperity, it would seem, therefore, that New York city acts as an exemplar of why a combination of strong territorial and network assets generates (advertising) work. Ultimately this interaction lies behind the development of a successful world city.

7 Los Angeles

A paradoxically 'local' creative city

> Los Angeles was one of the critical pulses of the economic and cultural condition of the Twentieth Century, and remains so.
>
> (Scott 2000: 171)

> LA is the centre of the entertainment industry and you can get almost anything you want here. You know, you have the best talent, you have the best production companies, you have the best scenery.
>
> (Interviewee 24)

Los Angeles' advertising industry, collocated with the global motion picture business, would seem to be primed to benefit from globalization. It might be expected, for example, that the Los Angeles offices of global agencies would act as leaders of global campaigns for major motion picture companies. In this chapter we explore whether this is the case by considering the forces driving Los Angeles' role in advertising globalization and specifically the territorial and network assets of the city. Using our data we reveal that, surprisingly, the cooperation between advertising agencies and the entertainment and motion picture industry appears to be much weaker than expected, the underlying reason being that the approach of filmmakers to advertising is different from other producers of consumer goods. In fact, surprisingly, the data we use shows that automobile clients feature very strongly within Los Angeles agencies' portfolios. However, Los Angeles demarcates itself strongly from Detroit as these clients are predominantly Japanese car manufacturers such as Toyota and Nissan. Indeed, one of the main ideas put forward in this chapter is that Los Angeles is less 'global' than might be expected, its role being as the 'capital' of the 'west coast' consumer market with networks reaching out to the Pacific-Rim and east Asia as a result of the desire of producers of consumer goods to reach this extraordinarily large market. As a result the city has gained from the increasing decentralization of advertising control from New York as 'cultural proximity' to west coast consumers is sought. Again we interpret these findings in the context of the work on the organization of advertising agencies, transnational projects, agglomeration and localization economies and project ecologies reviewed in Chapters 2 and 3, flagging the way 'west coast' consumer markets act as territorial assets that generate important network assets for Los Angeles.

Los Angeles: creative milieu and advertising hotspot

Advertising agencies have long been an important constituent of the Los Angeles economy. As far back as 1962, there were 506 advertising establishments (Standard Industrial Classification [SIC] 731) in the Los Angeles County, employing 6575 workers, with the average establishment size being thirteen persons (US Department of Commerce, County Business Patterns, quoted in Scott 1996). The contemporary Los Angeles advertising industry is a critical component of the city's cultural industries, which include fashion, digital media, architecture, entertainment and the arts (Los Angeles County Economic Development Corporation 2009).

Table 7.1 summarizes a number of key statistics in order to emphasize the significance of the advertising industry in Los Angeles in the early twenty-first century. The dominant employment size of establishments in Los Angeles is between one and four (397 establishments, 58 per cent share), and five and nine employees (108, 16 per cent) (US Census Bureau 2007), suggesting that, as already noted in earlier chapters, that the large offices of global agencies employing hundreds of executives are exceptional rather than representative of the average agency (such agencies represent only 2.6 per cent of all agencies in Los Angeles).

The growth of the Los Angeles advertising industry has been due, at one level, to the expansion and birth of Los Angeles-centred independent small and medium-sized enterprises, which form a highly competitive independent 'boutique' sector. Firms like Rubin Poster & Associates (RPA), David and Goliath and Quigley-Simpson Advertising have significant turnovers with, for example, the Santa Monica based RPA posting LA County billings of $1.141 million at the end of 2007 (*Los Angeles Business Journal*, quoted in Kyser *et al.* 2010). At another level the expansion of the top global and US nationwide firms in Los Angeles has complemented the boutique sections. By the mid 2000s, the city had representations from all of the leading global advertising groups although signi-

Table 7.1 Summary statistics relating to Los Angeles' advertising industry in the twenty-first century

Billings 2001 (US$ millions)	8,225.90
Global rank by billings (2001)	7
Number of headquarters of top six holding companies	0
Number of top six holding groups with one or more agencies in Los Angeles	6
Number of advertising agencies (NAICS 54181) 2001/08	735/718
Total employment in advertising agencies (NAICS 54181) 2001/08	11,362/12,641

Source: analyses presented in Chapter 4 and US Bureau of Labor Statistics Metropolitan Area Occupational Employment and Wage Estimates, Los Angeles County.

ficantly, a number of key agencies were absent from the city, instead choosing Irvine and/or San Francisco (for example, Draft/FCB, Young & Rubicam) as their Californian home(s). Despite this, Los Angeles is undoubtedly the most important advertising city in the west of the USA.

The significance of the advertising industry in Los Angeles today and in the past is no surprise given the historical and contemporary global reach and 'command and control' qualities of the city's industrial complex, initially because of advanced manufacturing in the city (for example aircraft production) and more recently because of high technology and services (such as finance) (Soja *et al.* 1983; Scott 1996). However, the story of Los Angeles' advertising industry is not the same as New York's industry with place-specific contingencies determining the role of global agencies' offices operating in the city. In the rest of this chapter we review, first, how the city's territorial assets explain the story of Los Angeles' contemporary advertising industry before, second, considering how the city's network assets are generated by and reinforce territorial assets.

Agglomeration and localization in Los Angeles' advertising industry

Unsurprisingly considering Los Angeles' reputation for clusters of cultural industries (see Chapter 3; Scott 1996, 2000), one of the major assets of the city is well-defined advertising agglomeration and localization economies. The advertising cluster has been identified by Scott (1996, 2000) as being a generalized, closely knit and distinctive district gravitating around Century City, west of Hollywood and east of Santa Monica. More specifically Kyser *et al.* (2010) make reference to the industry being centred in proximity to the 'Wiltshire' corridor, on or around the Wiltshire Boulevard from downtown central Los Angeles to Beverly Hills and Santa Monica. Figure 7.1 maps the locations of the offices of key global agencies in Los Angeles. This suggest that whilst there is a clustering effect as Scott and Kyser *et al.* suggest, for global agencies the cluster and associated project ecology is the greater Los Angeles city region rather than one area of the city.

Scott (1996: 308) suggests that the locational clustering of the advertising industry within Los Angeles can be accounted for by three interrelated tendencies which are generic to many other cultural products:

> First, the clustering of inter-related economic activities increases the (static) efficiency of transactions and information exchange between producers … Second, once these static effects have been secure, a more dynamic set of processes then come into play. These revolve around learning and innovation … existing as an 'atmosphere' of agglomeration-specific information and accumulated experience … Third, levels of economic competition … are often intensive, this maintains high levels of excellence.

The interviews we completed highlighted, in particular, the importance of the agglomeration and localization economies in Los Angeles for labour processes

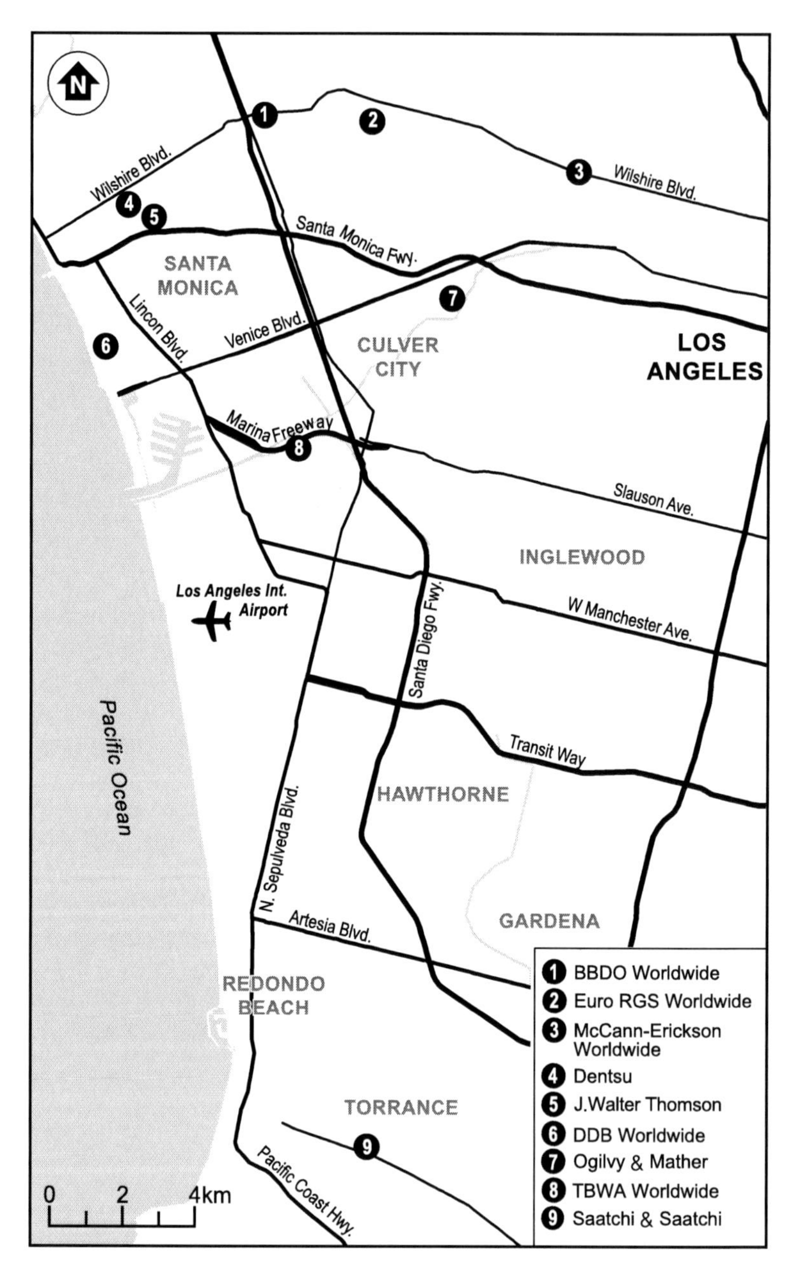

Figure 7.1 The location of key global agencies in the greater Los Angeles region (source: authors' research).

and practices relating to creative and strategic workers. For agencies, the creative labour force present in Los Angeles, which can be appropriated both through permanent employment of individuals by an agency and through exploitation of urban project ecologies which allow talent to be brought into temporary teams as and when needed, is crucial for knowledge acquisition, function, performance and brand projection. Put simply, to be successful agencies need to be able to produce advertising tailored to the Californian scene, something which requires talented strategic and creative workers. As one interviewee noted, 'the creative worker is always produced locally because you've ultimately got to be contextually relevant' (Interviewee 27).

However, in Los Angeles the labour market is tight for truly inspirational creative workers. As one advertiser described, 'the issue is that there's not enough creative people … in terms of number, the sheer universe of people to recruit from is actually quite small' (Interviewee 22). Some firms recruit from the local art colleges, but the majority of recruitment is generated from local labour market churn as creatives move between the array of advertising, media and creative industries in the city. For example, one account planner noted that:

> our churn of employees is about 25% … but LA is a city of opportunities … we always try to have talent in the pipeline … the very top creative people we … recruit nationally from places like New York … it's tough to get the senior more experienced people locally … [but] my guess is that about 80% of our recruitment is local.
>
> (Interviewee 27)

Nonetheless, despite the perception of a dearth of talent, in reality Los Angeles provides agencies with benefits in terms of recruiting workers that other US cities lack. Like New York, Los Angeles has reputational advantage as a 'creative' city and is well positioned to compete in a 'global' market place for creative industries such as advertising. This attracts a supply of labour from elsewhere within the US (for example, New York, Chicago, San Francisco) and from Europe (for example, London). Los Angeles is, then, a creativity magnet, sucking creative people into the incumbent labour pools of the advertising and wider, multi-media and design industries. The interviewee quoted above also noted that,

> very few people move to Detroit … I mean a lot of people go to New York, a lot of people move to London … Young kids … want to go and live in New York, Chicago or LA … These are the big diverse metropolitan cities, they're vibrant, they act as magnets to the affluent or educated professionals.
>
> (Interviewee 27)

Indeed, so powerful is the reputation of the cultural industries in Los Angeles that advertising workers often have a specific desire to work in the city because of the reputation agencies have for supreme creativity and innovation, and the benefits that working for a Los Angeles agency can bring in terms of future job prospects. As one account manager explained:

> they are actually going and joining agencies for the culture, for the opportunity, for the community that it is. I think that that actually is more prevalent in New York (and I say that because of living there), and probably Chicago because the industry there are many options and the community is somewhat centralized.
>
> (Interviewee 29)

In part the ability to attract talent and the effectiveness of the project ecology results from the wider array of creative industries in Los Angeles, not least the film industry. Table 7.2 details levels of employment in Los Angeles in key industries that make up the advertising project ecology. As described in Chapter 3, the project ecologies of the advertising and the motion picture industries overlap, both require, for example, providers of scenery and props for film shoots. Consequently it is the critical mass of creative industries that makes the city region such an attractive site for locating an advertising agency. Specifically the wide range of talented workers in the city region gives agencies a competitive advantage. As one interviewee suggested:

> the thing that drives ... [creativity] is the intellectual property. It's the ability that certain people have to be able to put these things together in a cogent fashion in 30 or 15 seconds and that's not something that can be stamp pressed out and turned over in Shanghai.
>
> (Interviewee 29)

There is little doubt that Los Angeles has a significant territorial asset in the shape of its labour pools. This attracts agencies to the city and, in turn, attracts clients to the agencies. But, this is not the only territorial asset that makes the city such an important site in the geographies of global advertising. The demand for

Table 7.2 Employment in 2008 in Los Angeles in key industries making up the advertising project ecology

Actors (NAICS 272011)	12,900
Graphic design (NAICS 271024)	10,710
Photography (NAICS 274021)	1,760
Film production (NAICS 274031)	3,180
Musicians (NAICS 272042)	3,320
Set designers (NAICS 271027)	940
Post-production editing (photography, music and film) (NAICS 274032)	5,460
Lawyers (NAICS 231011)	23,900

Source: US Bureau of Labor Statistics May 2008 Metropolitan and Nonmetropolitan Area Occupational Employment and Wage Estimates, Los Angeles County.

advertising agencies' services in post-industrial Los Angeles has also been driven by the yearning of producers of consumer goods for insights into the southern Californian and Los Angeles' 'anything goes' culture and the associated aesthetics, artistic distinctiveness, design subtlety and lifestyle. These insights are sought so that they can be used to inform the branding and advertising of goods and services in the USA, but also on some occasions in international markets (Molotch 1998). In short, these clients want to understand as Molotch (1998: 225) suggests the 'LA as design product', where the Los Angeles, 'local aesthetics ... affect what businesses produce and market' (also see Scott 2004).

Equally significant is the fact that it is not uncommon for the Californian economy to be compared with national economies because of its size. Indeed, California would be the eighth largest economy in the world if it were a nation state. And in many ways it is a distinctive economy, particularly when compared to the rest of the USA, something that means the identities of reflexive consumers and their spatial entanglements need to be catered for in advertising designed to run in west coast America. For advertising agencies being in the city is, therefore, advantageous because of the insights into consumer values, norms, attitudes and practices gained from sharing the 'lifeworlds' (Aspers 2010) of different groups of consumers. This forms a key territorial asset of Los Angeles. For producers of consumer goods, communicating with the California audience, which as already noted has more purchasing power than most countries in their entirety, through effective advertising is vital and, therefore, if advertisers can demonstrate knowledge of these consumers they have a competitive advantage.

Developing effective advertising for the Californian consumer means recognizing, as interviewees noted, that the identities of west coast consumers are starkly different from east coast US consumers. Thus, one advertiser in Los Angeles made the following comment about the importance of 'being there' in the city:

> drive around the streets in Detroit and tell me how many Toyotas you see. And Toyota is the number one selling car in America. And so if every day you don't see these cars on the street, how can really understand the world you're operating in?
>
> (Interviewee 27)

This interviewee also went on to highlight how Californian culture is itself heterogeneous. Most notably Los Angeles is a crucible in which Hispanic and white American cultures 'collide' and hence:

> In some examples, what works in Florida with the profile of Florida is very different from what works in Southern California because we have immigration, we have a huge Hispanic population in Southern California and there are Hispanics from South America here whereas the Hispanic population in Florida is much more Cuban and Caribbean based and so there's different behaviours and different motivations.
>
> (Interviewee 27)

An additional territorial asset of Los Angeles is, then, the access it provides to key consumer markets. However, considering the size of the Californian consumer market, it is surprising that, for the global agencies we studied, alongside access to a large and diverse consumer market, a large pool of potential clients is not one of the territorial assets of Los Angeles. First, our research revealed that these Los Angeles based agencies had very little cooperation and linkages with the city's global motion picture industry. This was due in most part to the different approach that film companies and their distributors have to advertising their products through both digital and printed media platforms. As one interviewee noted:

> When they launch a movie they're in the market for two weeks. I used to know people at Fox studios. They just spend a ton of money over two weeks and then moved onto the next movie. And all their money is spent on Thursday and Friday because they want to get people to go to movies Friday and Saturday night whereas here if I launch a car, it's got to be successful for five years.
>
> (Interviewee 27)

As noted in Chapter 2, global agencies sell their services based on the long-term brand stewardship they can offer. Short-term campaigns are not, therefore, something agencies regularly seek to get involved in. Consequently, most advertising agencies, including the boutique agencies in Los Angeles, do not view the motion picture industry as an important source of work. As another interviewee described:

> the dynamics are a bit different. The studios really work with media placement agencies. Western International would be one of them. Sony or Buena Vista had just found that they manage the creative, that's one of the most important things you pointed out like trailers, since they really control the creative by and large they don't necessarily need the agency to do anything to place it. The irony of this is can the secrets of the business, Sony ... Universal, all these people will out-source their creative, so even though they've got these people who will do the trailers, the trailers are actually done by agencies. Different types of agencies, they're movie agencies but they specialize in creating trailers. There are also agencies who specialize in doing the advertising with that and create it, most of its very small; 3, 4, 10 people something like that and may just be contracting project based type stuff, but by and large almost all of that is outsourced.
>
> (Interviewee 29)

As such there is a fundamental culture clash between the philosophy of many advertising agencies and the motion picture industry in terms of advertising strategies because:

> It's very hard when you're dealing with people who've created a two hour film and the trailers and everything else, it's very hard indeed for them to take advice on how to create something from an ad agency whereas if you're a Procter and Gamble selling soap powder, you're not particularly creative. So it's tough. And the people that run the studios here are extremely egotistical, entrepreneurial, strong-willed people. It's very hard for them to take any advice on anything. What you get is a lot of media placement and maybe just putting the thing together but the creative gene, no they think they've got it sorted.
>
> (Interviewee 30)

The result is that, 'there are two advertising businesses in LA. There's advertising of consumer products which is what we do, and there's the advertising of films' (Interviewee 22). This is not to say agencies do not work for any 'local' clients. But, local client bases do not form one of the territorial assets that attract agencies to Los Angeles. Rather it is clients accrued through network assets which are themselves generated by the territorial assets of the city that are of most importance.

Globalized advertising networks: reasserting the local

The international gravitas of the Los Angeles based agencies we studied is clearly demonstrated by their 'blue-chip' clientele, which include major global firms and brands such as Nissan, Apple, Playstation, Visa, Nestlé, Land Rover, Aston Martin, Hilton Hotels, Toyota, Procter & Gamble, Lexus, Ritz Carlton Hotels, Cisco Computers and Singapore International Airlines. In particular our research confirmed that Los Angeles holds the position of the 'New York of the Pacific Rim' (Soja *et al.* 1983: 211). Being the New York of the Pacific Rim means being the landing point for Asian producers of consumer goods, especially automobile manufacturers like Nissan, Honda, Toyota, Mazda, Subaru and Suzuki (Scott 2000).

Of great significance is the fact that the Los Angeles offices of global agencies are often responsible for representing the entirety of their overseas clients' USA advertising business activities, which means leading collaboration and cooperation between multiple US offices to deliver campaigns. In some cases, the Los Angeles offices of global agencies have an even wider remit, being the lead office for campaigns running inside, but also outside the USA, for example in South America and Asia. This means leading collaboration and cooperation between offices in cities such as Sao Paulo and Singapore. Table 7.3 captures these network assets and their positioning of Los Angeles in geographies of advertising globalization.

In order to understand these network assets, it is important to again recognize the role of Los Angeles' advertising industry in servicing the gigantic Californian consumer economy. As one advertiser suggested,

> the reasons why agencies landed here had more to do with big clients that actually were here … but, not entertainment … a lot of the bigger agencies

Table 7.3 The connectivity of Los Angeles in terms of advertising work

Ranking of connectivity via office networks of global agencies	0.44 (New York ranks 1.0)
Score of importance of the city in terms of connectivities associated with European advertising work	–0.07 (i.e. insignificant)
Score of importance of the city in terms of connectivities associated with Asian advertising work	1.23 (0.13 ahead of New York)
Number of headquarters of ten largest global agencies	0

Source: analyses presented in Chapter 4.

> were frankly here because of the automobile clients … LA was made by the import clients.
>
> (Interviewee 23)

And the rationale for these clients using Los Angeles agencies to develop advertising for the Californian economy was simple. As two advertisers put it: 'When I look at Nissan and Infinity, I would say … we handle North American business, we also work with them on a global level … There was a belief in this sort of closeness' (Interviewee 23), and 'Everything is competitive … we'll do extensive campaigns and we have done … work for …Toyota … very, very and specific local … what ends up happening is that business is very much regionalized, it is very local' (Interviewee 29).

The Los Angeles offices of global agencies have, then, become the 'core or hub of ideas' (Interviewee 23) for many non-US producers of consumer goods wishing to engage west coast US clients. The deep understanding of the Californian, west coast genre or consciousness that advertisers in Los Angeles possess, gained by sharing the 'lifeworld' of consumers, means an understanding of the rhythms of the city of Los Angeles and west coast lifestyles can be fed into account planning and creative decision making. The following vignettes from our interviewees illustrate the relevance of the argument that Los Angeles' place in global advertising networks is in significant ways built around the development for global clients of advertising for 'local' markets:

> if we're pitching a LA based client, they generally want LA people working on the business … more often than not we go into a pitch in California and the client will say, 'do all these people live in this town?' … one of the criteria would be that you have an office in town and its run by people in town because they want you in the same time zone, they want to see the creative people, they want to see the agency.
>
> (Interviewee 24)

> I can tell you the difference between New York and LA. In New York the joke is 'you are what you wear' and in LA 'you are what you drive'. In New

York, they don't care about what they drive ... Here people have three or four cars: they'll want their week-end car, their go to work car because you're not making as bigger economical legislative sacrifice to own a vehicle. So it becomes a little more status driven, a little more intangible, a little more ... frivolous.

(Interviewee 27)

But, it is not just adverts for west coast US markets that the Los Angeles offices of global agencies develop (see Table 7.3). Working for global clients on occasions also involves meeting their needs across the USA. As one interviewee surmised, 'we do work for a global client ... [We are] still doing purely local work ... on a brand like [brand x] ... but the role of the LA office ... [is also] to do the work for the US market' (Interviewee 26). One agency's work for a major hotel client illustrates the spatial division of labour and organizational networks within the US that the Los Angeles office is part of:

> we work very closely, with the Chicago office. They do the creative work on [client x], we do all the strategy and account service. We've partnered with San Francisco on some things. If our creative department isn't really cracking something, we'll ask New York creatives to help out and they've asked us to help out but we don't always work on everything together.
>
> (Interviewee 30)

The two way collaborations described by this interviewee are vital as they both generate significant work for the Los Angeles offices of agencies when they lead campaigns and ensure the adverts produced by the Los Angeles office are of the highest quality thanks to assistance with campaigns given by other US offices. Particularly when campaigns are run using the transnational model within the USA – i.e. when multiple campaigns are developed for different regions of the country – either the leadership role played by the Los Angeles office or its collaboration and cooperation with other offices generates work that would not exist without such network assets.

In addition, the Los Angeles offices of global agencies also have network assets in terms of their work on campaigns outside the USA (see Table 6.3). First, many agencies are engaged in global work from the Los Angeles office in a leadership role. For example, two advertisers suggested that:

> 20% of the staff work on global things ... we can't not be a part of the global conversation, especially in advertising ... we are part of a very strong global networks, we do have global clients ... *I can now tap into the global city network.* I'm much more likely to talk to somebody in London or Tokyo or Paris as I am to call somebody down the corridor.
>
> (Interviewee 23, emphasis added)

> [Client x] employ us as a network. We lead the work for them in Japan, Great Britain and Europe ... on a regular basis four or five times a year we

> meet with other offices ... to share learnings, share applications, compare notes, make sure anything new, different or successful ... are shared.
>
> (Interviewee 27)

Second, on other accounts the Los Angeles offices cooperate and collaborate with leadership provided in another office overseas. This means, for example, being part of multi-office teams to pitch for new business with existing or new global clients and brands. As one interviewee summarized:

> if we are pitching a [client x] worldwide assignment the meeting would be in [European office x] and so in preparing for that meeting we'll assemble a team of account and creative people from the USA, from Europe, from Asia-Pacific, from the Far East, from around the world. So there will be a group of account and planners and creative groups that will work on it, they'll probably work independently on the assignment then they'll go to [office x] and get together and spend the last week working on it together and then present it ... The other way it works is say, we needed help on a brand, we would ask [office x] to develop some creative ideas for us. They would either send them out or they would come out we would put it together with the work that we have developed and generated and then we would go to the client and present it. So you know it can work it can go it can working globally or working internationally or supporting other officers can happen in a number of different ways depending on the assignment.
>
> (Interviewee 24)

As this quotation suggests, just like intra-US collaboration and cooperation, membership of global account teams generates work for the Los Angeles office and helps support the quality of the adverts produced by the office. The fact that the nationality of advertisers in Los Angeles included US nationals, but also Brazilian, British, Chinese, Mexican and Singaporean executives reveals another benefit to such network assets: the ability to tap global talent pools. The international makeup of the offices allows the skills and competencies of overseas workers to successfully dovetail with the Los Angeles and 'west' coast locales who have the vital in-depth knowledge of Californian consumers discussed above.

Third, agencies in Los Angeles increasingly draw on expertise from throughout their firm's office network to generate work that is executed in Los Angeles. One commentary from an interviewee which fully elaborates the nature of this network based work generation is particularly insightful:

> I think that what we're proving as a company the [agency x] network of companies can work anywhere. We just won [client x] out of our Singapore office ... The reason I'm telling you that is because our Singapore office won the business ... We didn't pitch anything ... One way to look at the future is we will dispatch locally, but we will present the bandwidth and firepower of a global enterprise. That speaks to the future that we can take on these different

> projects because we've got a methodology that's universal and proven, and we have a resource of different people who can help make it happen ... So I think that's the future. For us to jump out and say 'well, we wanna Taiwanese manufacturing companies out of LA', I would never do that anymore. I would say 'who's our Taiwanese office? Great, let's work with the Taiwanese guy, let's put together a package and let's pitch it together'. I think it's more of a win–win because the local people get to win and we get to win because we're gonna work with the local people and provide some firepower and some expertise, and so it's gonna work both ways. The other point I'd like to make is that typically, the Singapore office would wait for the LA office to have a project. And here's something where the Singapore office led it, drove it and brought us in as needed. They're gonna own that piece of business and we're gonna serve them, and I think that's a much more virtuous model.
>
> (Interviewee 22)

China was identified by several agencies, who already have an office presence in the country (in Beijing, Shanghai, Hong Kong), as a gigantic potential market for their clients, especially the automobile companies, and hence a potential market from work might also be generated.

Finally, and unique to Los Angeles, global network assets were also generated through the role of the office in actually shooting advertising campaigns. Often by linking with international offices, Los Angeles could generate work thanks to the city's deep pools of talent and sophisticated project ecology, which as already noted is partly generated by the benefits of collocation with the motion picture industry. This reveals the importance of the co-presence of other industry clusters in a city in terms of the development of effective localization economies which can themselves generate network assets.

Implications

Both the territorial and network assets of Los Angeles play a multitude of roles in making the city an important site in the geographies of advertising globalization. Territorially Los Angeles benefits both from the proximity it has to key consumer markets and the city's deep pools of labour and sophisticated project ecology, reflecting the New York story. The importance of the west coast consumer market as a territorial asset reminds us of the US exceptionalism discussed in Chapter 4 and explains this with reference to the heterogeneity of consumer identities and entanglements in different regions of the country. Indeed, at one level it could be said that the cause of Los Angeles' central role in the advertising space economy requires a much more local explanation than one would expect for the 'movie capital' of the world. Two substantive factors can be distilled from the research in Los Angeles which accounts for the agencies' disproportionate concentrations of *local work* in the city and on the west coast: agencies' disconnection from the highly globalized motion picture business, and agencies' reliance on global corporate clients, most notably Japanese and European automobile manufacturers, as well as US clients and their need for advertising that targets west coast US consumers who buy into the Californian genre and 'lifestyle'.

However, if read as solely a 'local' advertising city, the work of the global agencies we studied in Los Angeles seems somewhat fragile and potentially under threat in the future. Each of the Los Angeles agencies recognized that the future would bring different challenges and opportunities for their operations on the west coast, with all agreeing that the most immediate and prevalent issue was competition from other agencies in the city, especially the smaller and more specialized 'boutiques'. As two interviewees suggested:

> the enemies of this company are not within the walls of this company. The enemies are on the outside. A company has to realize that … 'enemies' sounds quite warlike, but in our business, it is very cut-throat and we are here to service these clients.
>
> (Interviewee 22)

> Big agencies … are more afraid of the small start-ups from the US than they are afraid of [global agency x] or [global agency y] or any of the big companies around the world.
>
> (Interviewee 28)

Los Angeles, as we have already discussed, is an incubator for independent agencies, which often specialize in particular functions (e.g. media, marketing, digital, health care). Boutiques like, Rubin Poster & Associates, David and Goliath, Quigley-Simpson, are making significant inroads into large agencies' business and have successfully pitched for business with major corporate clients such as Honda and MGM, Kia Motors and Mattel, and Proctor & Gamble and Visa respectively (www.allbusiness.com, accessed 1 April 2010). In particular, small boutique agencies are increasingly seen as more flexible, creative advertising organizations in contrast to global agencies which have been accused of being too bureaucratic and slow to adapt to new client demands.

However, the role of Los Angeles in advertising globalization should not simply be read as being exclusively 'local' and tied to west coast US consumer markets. This does not mean global agencies should not be worried about the threat from boutique agencies. But, it does mean that global agencies can benefit from global work in important ways and differentiate themselves from many boutique agencies. The territorial assets of Los Angeles – large consumer markets, deep pools of skilled labour and a sophisticated project ecology – generate important network assets in the form of leadership and collaboration and cooperation on intra-US and global campaigns, resulting in levels of advertising work in the city that are much higher than would otherwise be expected. Figure 7.2 captures these interdependencies between territorial and network assets, interdependencies that are similar but subtly different from those in relation to New York. This reminds us of the importance of the geographical contingency of any city's role in advertising or any other processes of globalization.

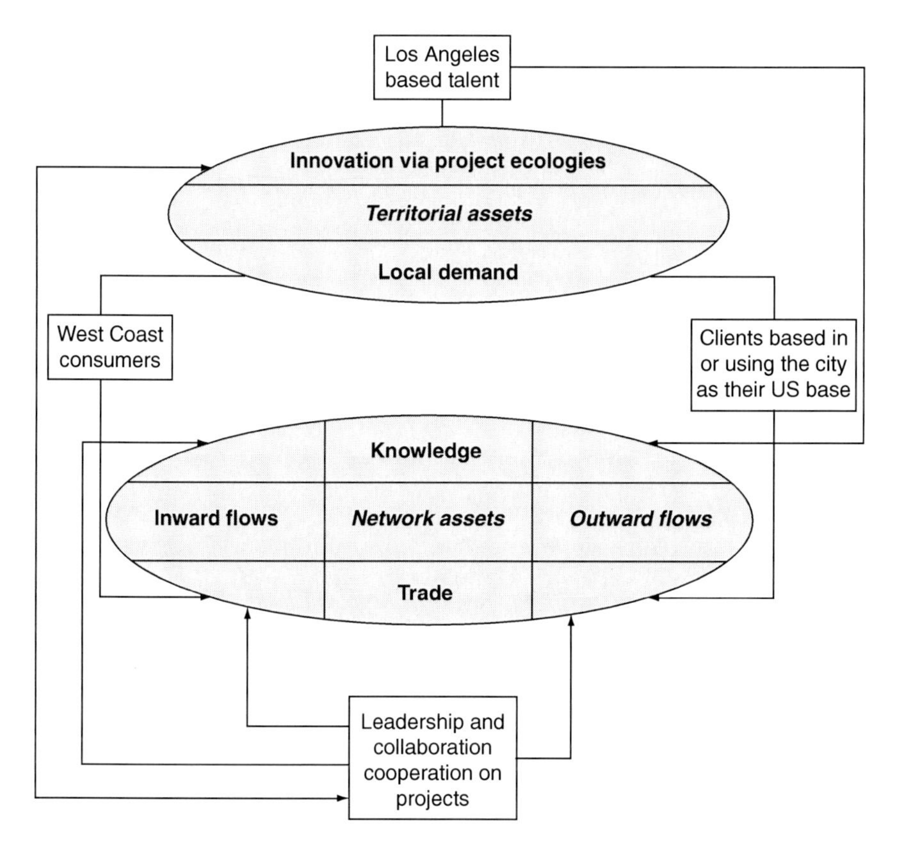

Figure 7.2 The territorial and network assets of Los Angeles and the interrelated and inseparable ways that they make the city an important site in the geographies of advertising globalization.

Conclusions

Los Angeles is an example of an advertising city that can credit its success in terms of levels of employment and work to a combination of place-specific territorial assets and related and interdependent network assets that generate global work. In this chapter we have explored the way global agencies have developed a niche in Los Angeles that does not necessarily involve working for the motion picture industry, but instead involves developing campaigns for west coast US consumers whilst also leading worldwide campaigns for key clients operating in and through the city. As such Los Angeles provides another example of how the theoretical framework developed in the first half of the book can be applied to a specific city to explain its role in advertising globalization whilst also further helping explain the US and worldwide geographies of advertising globalization reported in Chapter 4.

8 Detroit

Market change and a city falling outside the global space economy

For many Americans, Detroit is Ground Zero.

(Luther Keith, Executive Director of Arise Detroit, quoted in Popelard and Vannier 2010)

The historical implantation in the twentieth century of the 'Big Three' motor manufacturers (General Motors, Ford Motor Co. and Chrysler) in Detroit unsurprisingly turned the city into one of the major centres for the US advertising industry as many of the leading global agencies set up offices in the city to service the accounts of these giants. At the turn of the Millennium this led to Detroit being in seventh position in the list of US advertising cities with billings reaching US$8336.80 million, as shown in Chapter 4. Some of the major global advertising agencies operating in Metropolitan Detroit at the turn of this century included Leo Burnett, Campbell-Ewald, J. Walter Thomson, Young & Rubicam, Olgilvy & Mather and BBDO as well as a number of large independent agencies such as Doner and Mars Advertising.

Detroit's role in the geography of the US advertising industry, and the geographies of advertising globalization more broadly, reflects its specificity as a single industry town with a few large and powerful clients in the shape of automobile manufacturers. This has, however, led to a downwards spiral in which Detroit has been stuck over recent years as a result of its failure to develop a post-industrial urban economy to replace continually falling levels of manufacturing employment. To understand the position of Detroit in the US and on the global map of advertising, this chapter considers the territorial and network assets of the city, focusing in particular on the strong dependency that its advertising agencies have on the automobile industry, the specificities of car advertising in relation to consumer expectations and preferences, Detroit's ability to develop advertising that addresses these expectations and preferences, and the way the special relationship with automotive clients shapes the nature of the creative and strategic work conducted by Detroit agencies and the network assets associated with global work.

A picture of Detroit: a ruin of a city?

Detroit is undoubtedly the US city that has the most strongly demonstrated the benefits and costs of urban rise, decline and renewal. A shrinking (and even 'Pariah') city par excellence in the twenty-first century, Detroit has captured the attention of numerous planners and sociologists who have sought to decipher the contours and processes of urban sprawl, racial residential segregation, polarization and poverty (see for example, Kodras 1997; Sugrue 2005; Wilson 1992; Wilson 1999). A plethora of core city regeneration projects through business, arts, entertainment, community and tourism have also been studied since the early 1990s. But the city's abandoned skyscrapers are clearly a reflection of a failed Renaissance (see for example, Allard *et al.* 2003; Eisinger 2003; McCarthy 1997; Neill *et al.* 1995; Steinmetz 2009).

It is worth bearing in mind that Detroit, as the largest city in the State of Michigan, has existed since 1701. Its population rocketed from just 116,340 to almost a million between 1880 and 1920, making it the third largest US city after New York and Chicago. In the nineteenth century, its industrial base was diverse, but between 1909 and 1914, 85 per cent of the national growth in the automobile industry occurred in Detroit and between 1914 and 1937, 47 per cent of the total US workforce employed in this sector was located in Detroit (Steinmetz 2009: 762). In 1913, Henry Ford and his assembly-line method for making the Model T became an inspiration for the new industrial age while Detroit's Jefferson Avenue Assembly Plant rolled out its first Chrysler in 1925. The city's automobile technology spread throughout the world and the big luxurious American automobile became the epitome of US know-how. Detroit's population increased more than sixfold during the first half of the twentieth century, fed largely by an influx of European, Lebanese and southern migrants to work in the burgeoning automobile industry which consolidated its position after the Second World War. From then on, the process of urban sprawl started to unfold as the affluent automobile workers moved out to settle in the wealthy suburbs.

However, by the turn of the twenty-first century Detroit, the city supposed to represent the American Dream, was in crisis. According to the US Bureau of Labor Statistics, unemployment for the Detroit Metropolitan area was above 7 per cent when we conducted our research in 2007 and has since steadily risen to over 16 per cent as a result of various crises facing the 'Big Three' auto manufacturers. In particular, competition from Japanese auto manufacturers has slowly eroded the dominance of the 'Big Three', reducing their share of the market for new cars and thus production levels in Detroit. Moreover, the cyclical nature of the automobile industry, prone to booms and busts as demand rises and falls with economic cycles, has further made the city vulnerable to shrinkage and out migration. Indeed, according to the US Census Bureau, between 1990 and 2000, Detroit's population shrunk by over 10 per cent from 1,027,946 to 916,936 with declines continuing to the present day.

So what does all this mean in relation to our discussion of advertising globalization? To understand the changing nature of work of global advertising agencies in Detroit, it is essential to understand this historical and contemporary substratum upon which Detroit's advertising industry is built. In particular, this substratum affects the key territorial and network assets of the city.

Advertising agencies and Detroit: flimsy territorial assets

Table 8.1 provides summary data about advertising work in Detroit in the twenty-first century, showing that perhaps surprisingly the city's level of billings exceeded those of Los Angeles with only New York, Tokyo, London, Chicago and Paris generating more billings. However, the story of advertising in Detroit is unlike the stories told thus far through the cases of New York City and Los Angeles, the rapid decline in employment in advertising between 2001 and 2008 being the most obvious evidence of Detroit's distinctive story.

The Detroit Metropolitan Region is home to many agencies, but, strikingly, none of the large advertising agencies are located in downtown Detroit. The agency J. Walter Thomson was located in the city centre in the heydays of Fordist production, but has since moved to suburban Michigan, closely mirroring the location of the Big Three auto manufacturers. The city of Dearborn (Wayne County), the tenth largest city in the State of Michigan and home to Ford and the WPP group agencies that service it, is also an important location for agencies. Other agencies that service Chrysler and General Motors can be found in Warren (Macomb County), Detroit's largest suburb and business centre. Warren is also home to Campbell-Ewald, which has handled GM's Chevrolet work since 1922. Figure 8.1 shows the location of the global agencies with presence in the Detroit Metropolitan region, confirming the agencies' dispersed locational strategies.

The fact that agencies are not located in downtown Detroit but are spread across the metropolitan area can be explained by the Detroit region's unique territorial asset: the presence of the 'Big Three' automobile manufacturers and the relationships between agencies and these manufacturers. Each agency has located its offices in proximity to the manufacturer it serves, with each agency only serving one of the manufacturers' brands because of conflict of interest issues. This does not, of course, preclude one holding company such as WPP

Table 8.1 Summary statistics relating to Detroit's advertising industry in the twenty-first century

Billings 2001 (US$ millions)	8,336.80
Global rank by billings (2001)	6
Number of headquarters of top six holding companies	0
Number of top six holding groups with one or more agencies in Detroit	4
Number of advertising agencies (NAICS 54181) 2001/08	283/228
Total employment in advertising agencies (NAICS 54181) 2001/08	5,786/4,638

Source: analyses presented in Chapter 4 and US Bureau of Labor Statistics Metropolitan Area Occupational Employment and Wage Estimates, Macomb, Oakland and Wayne (Detroit Metropolitan Area).

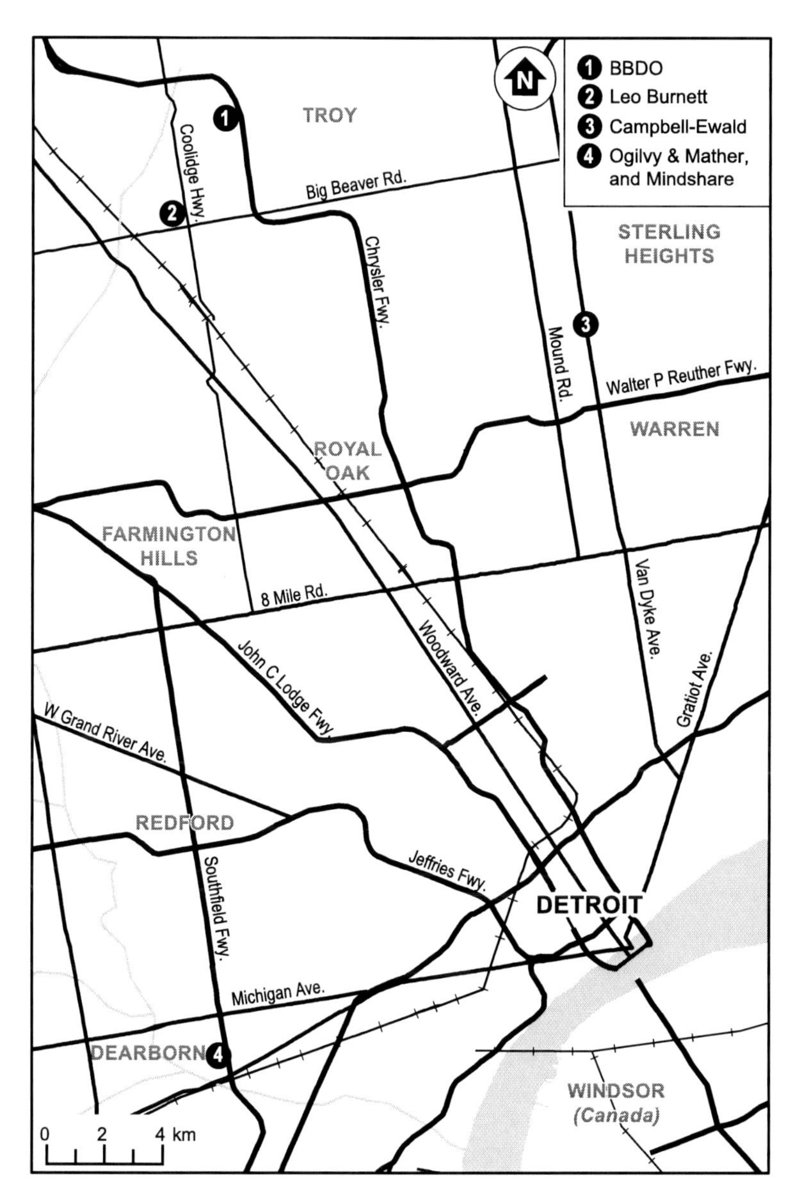

Figure 8.1 The location of key global agencies in the Detroit region (source: authors' research).

providing advertising for different brands through different agencies because of the 'Chinese walls' setup (see Chapter 2). Considering the close ties between agencies and particular auto manufacturers, it is unsurprising that one of the recurrent themes in our interviews was the extent to which agencies' relationships with the automobile industry dictated the type of advertising work done in

Detroit. In particular, because servicing the automobile industry was seen by interviewees as being extremely different from servicing other consumer products, especially disposable goods such as foods, beverages or toiletries, the unique strategy has been adopted of setting up agencies in Detroit dedicated to serving a single, automobile industry client. At one level this appears logical. Next to purchasing one's home, buying a car is both financially and emotionally one of the most important consumer purchases. Pitching for cars has very specific features because as opposed to other non-durable goods, the consumer's emotional bonding with the product is much more intense and needs to be targeted in particular ways. As one interviewee noted:

> Some people have very different relationships with beverages than they do with automobiles. We happen to think that the relationship with an automobile is so intense that you could probably find a similar trait throughout all people. If you're drawn to that vehicle, there's something that you and I might share even though we're 2000 miles apart. But maybe your relationship with water can be very different.
>
> (Interviewee 20)

Other reasons that make handling key global car accounts a complex task for advertising professionals include the fact that creative and strategic work does not only amount to making the car desirable through strategies focused on a limited number of TV or print advertisements. In the case of the automobile industry, advertising is also deployed by the client's distribution and repair network. This advertising needs to be developed to ensure image consistency in retail facilities, something which may involve creating specific adverts for individual dealers to use in their local newspapers or on radio or instructing them in how to tap into online social networking sites and develop 'viral' campaigns. In addition, one of the significant differences compared to other product categories is that the product's lifetime is longer. Agencies consequently benefit from longer planning cycles, typically eighteen months, which is much longer than with other goods. Indeed, even quick turnaround campaigns that involve shortened development times are still slower than many of the campaigns developed, for example, by agencies in New York. As one interviewee describes the timescales associated with advertising campaigns for cars:

> They're trying to shorten it [the product cycle]. We know what's going on in automotive out until the next 15 to 20 years. So we start on a project a year and a half in advance. A lot of agencies don't have the kind of longstanding relationship that we have. There's a lot of companies, they come in and the client says 'I have this new product and I need a campaign in 60 days' and so they have to figure it out in 60 days. We know so much going into it and we have a year and a half to try and plan for it and anticipate what's going on. But that just goes with having a long-term relationship.
>
> (Interviewee 20)

Hence it would seem to make sense to locate agencies in Detroit in proximity to automobile industry clients and allow agencies to manage campaigns almost exclusively for Detroit based auto clients. Three advertising executives, mirroring the comments of several others, noted that:

> We are an agency that was born out of [auto client x], we used to have all the business ... Detroit, because of the auto industry and the vast economic engine that it is was, has been a significant agency outpost and because the accounts are so big they have tended to have their leadership right here because if they were flying in from New York they would probably be feeling under-served. So we've had the [auto client x] account since it was founded over 90 years ago.
>
> (Interviewee 19)

> It still comes down to if the client who's writing the checks is located in Detroit – which is why there are such big agencies in Detroit – those clients want that person close by that they can work with. Is there ever going to be the chance that [auto client y] is going to want an agency out of Shanghai to run their business?
>
> (Interviewee 18)

> Pretty much 99 percent of what happens in terms of marketing and communications is from this office for [auto client z], which is rare because a lot of companies have moved to having an office doing their advertising, hiring another one to do their direct mail and another one to do their online.
>
> (Interviewee 20)

Detroit undoubtedly has, then, a key territorial asset in the shape of automobile industry clients that attract advertising agencies to the city and generate work. This explains the city's high ranking in tables of annual advertising billings and its success as an advertising city. However, the key territorial asset of the city is also a potential impediment, tying agencies into relationships that are flimsy, over reliant on one client and their economic success and advertising needs, and preventing a diverse portfolio of advertising work being developed because of the difficulties of managing automobile industry accounts and their peculiarities alongside other client accounts. This led, for example, to interviewees describing a dedication and even subordination to clients that was not experienced by agencies in other cities working for a more diverse array of clients. Such dedication and subordination may, for example, involve foregoing autonomy about the choice of members of a project team for a campaign, such as photographers, editors etc., with the client dictating the terms. Such abnormal relationships were accepted, though, because maintaining the special relationship between an agency and an auto manufacturer client is seen as vital. The revenues that can be generated by working for such clients are great: in 2009 the automobile industry spent over US$15 billion on advertising (*Advertising Age* 2010d).

One noteworthy example of this is case of WPP's 'Team Detroit', which pulls together resources from multiple shops to service a single large global account. Following a 'global pact' signed with Ford, Team Detroit was established in 2008 as a joint venture comprising WPP's five Detroit offices: JWT, Young & Rubicam, Ogilvy & Mather, Wunderman and Mindshare (with the latter two handling digital and direct marketing operations). Figure 8.1 shows how these agencies are collocated in the same building (Wunderman and Mindshare are not listed on the map, but operate in the same building as the three WPP agencies that are listed). The underlying idea is to provide Ford with a one-stop shop. This has clearly been a strategic move on the part of WPP's Chief Executive Sir Martin Sorrell to not only integrate services, but to change the network's culture which is reflected in the following statement on Team Detroit's website:

> we don't do it the old way. We do it the new way. We do it in a better way. Over 1,000 people joined together from every marketing and communications discipline to create iconic and effective work. We are one. We are Team Detroit.
>
> (see www.teamdetroit.com (accessed 1 September 2010))

Such investments by holding groups and agencies are, however, risky ventures. The data gathered from our interviews shows that relationships between agencies and clients in Detroit are characterized by fluctuations, depending on who is heading the brand's marketing and communications strategy at the client's headquarters. In addition, what comes out clearly from our interviews is that the senior sales and marketing executives also forcefully impose their visions for the brand. If the preferences of the client's marketing and communications executives change in relation to the image of a brand or the use of an agency, there is the potential that a relationship may be broken leaving Detroit based agencies effectively without work. As one advertiser noted:

> It's one of those double edge swords in that it's very risky but then also for a while [client x] were large enough that they could say 'we don't want you in this office to work on anything. We want you to focus totally on our business and have all the resources available'. Because of the size, you say 'okay'. Because then you have the office in Chicago that handles all the other diversified clients. But we've realized that in the last 6 months that we need to diversify and we need to start that process. It's hard though when you're seen as the 'car agency' to have some other category say you have an expertise they can capitalize on.
>
> (Interviewee 18)

Detroit's main territorial assets, a large auto industry client base and the benefits it brings in terms of generating advertising work, is also its Achilles heel. This is not just because agencies are overly specialized and subject to the whims of an individual client. Detroit's reliance on the automobile industry as a source of employment, literally in terms of those employed in the factories and laterally in terms of work generated in advertising agencies and other business services, means the city lacks deep and diverse labour pools, sophisticated project ecologies and large and diverse consumer audiences – i.e. key advertising territorial assets – when compared to other US cities, and in particular, New York and Los Angeles.

Specifically, in terms of labour pools and talent, at first glance, Detroit does seem to boast one territorial asset: what might be referred to as an 'institutional thickness', comprising a depth and quality of local governance networks in economic development (Amin and Thrift 1994). In the case of the Detroit advertising industry, institutional thickness relates to the trust, knowledge and memory the individual executives have built and stored over the years. This accumulation of knowledge proves invaluable when it comes to preparing pitches to clients. It makes it possible to save time and money, avoid errors made in the past and more efficiently develop effective campaigns that will please the client. Certainly, some of the Detroit based advertisers identified this memory as a significant asset:

> Our automotive service business, that hasn't gone in too many different places. It hasn't because our clients here want a knowledge base and we very much service their institutional memory. The turnover at large companies is fairly high. The turnover at this agency hasn't been so high. I can't tell you why that is. We service an institutional memory.
>
> (Interviewee 16)

> We used to be the [brand x] agency. We were the ones that turned [brand x] around. We used to spend a lot of time with our offices giving them the case history of what we did to save [brand x] and whenever we'd start a project, we would communicate all around the world to the [brand x] offices … That's one thing we consider really important and that we have here in this office. We have this institutional knowledge. There's so much knowledge here as [client x] clients come and go, this office is one that has all that history and knowledge and can fill in the blanks.
>
> (Interviewee 18)

> There's a large learning curve for other agencies that come and try to touch automotive. Most agencies that have been given an automotive account haven't fared all that well. The agencies that typically do well have some experience and that includes overseas as well partly because there's such a high learning curve. Out of the gate you might be able to do well but then, to be able to sustain the success is difficult. It's challenging.
>
> (Interviewee 20)

However, mirroring discussions of the danger of lock in caused by institutional thickness (Grabher 1993), Detroit's institutional thickness manifested in the labour pool can also be an impediment. Because of Detroit's industrial heritage and failure to emerge as a post-industrial city, the advertising workforce in the Midwest is often described by outsiders, and sometimes by insiders, as being insular, lacking creativity and unable to develop campaigns that generate the real emotional attachment to a product that is needed to make a campaign successful. Whilst they may be good at keeping clients happy, it is suggested that in doing so, the campaigns developed are often less effective than they might be. This problem in many ways relates to Detroit's lack of a diverse consumer audience, caused by high levels of unemployment in the city, and the relative homogeneity of the employed population because of the auto industry's domination of the regional economy. Thus, many advertising workers are seen as being unable to connect with consumers outside the Detroit and Michigan areas of the USA. As one interviewee suggested:

> Detroit I find in retrospect incredibly insular. I think that's a classic example where they assume that the rest of the world is like them and that's not the case. Drive around the streets in Detroit and tell me how many Toyotas you see. And Toyota is the number one selling car in America. And so if every day you don't see these cars on the street, how can really understand the world you're operating in?
>
> (Interviewee 27)

Indeed, in wider debates about the future of the advertising industry and the ineffectiveness of campaigns developed in cities throughout the USA, one advertising professional in a personal blog devoted to the future of the industry sarcastically used the title 'Ad Aged – Will Madison Avenue become Detroit?', the reference being supposed to allude to the dim view many advertisers have of the campaigns coming out of Detroit (see http://adaged.blogspot.com (accessed 1 September 2010)). As a result, Detroit's advertising professionals rarely switch to other fields and acquire a high degree of work specialization, highly valued by the local client, as already mentioned, but often viewed dubiously in the rest of the industry. In the words of an LA based interviewee:

> We call them car people. Car people are car people for life. But the car person is a real persona and I've had friends who have been hired by Detroit agencies at a large network because they've only ever been able to do car business.
>
> (Interviewee 29)

And the domination of the automobile industry in terms of employment in Detroit has another effect on the city and its territorial assets. The city has a weak milieu or project ecology because of the lack of other cultural industries with either related or unrelated variety. Table 8.2 details employment in industries that make up the advertising project ecology and reveals that levels of employment are extremely low. Indeed, when no data is available this is due to the low levels of employment in the occupation and an inability to generate representative data that the US Bureau of Labor Statistics is willing to release. As a result, agencies in Detroit find it extremely difficult to get hold of the talent they need to make up temporary project teams. An interviewee recalled this problem in the following way:

> There was a local meeting of creative directors in Detroit. The subject was 'we need to band together to make sure that the suppliers in Detroit remain viable financially so we need to band together, make sure we use them'. I said at the meeting: 'I don't feel like I have an obligation to use a supplier simply because it's here. I have an obligation to my client to provide them the best answer. If I think that's in Mumbai, it's in Mumbai. If I think that's in New York, in Cleveland Ohio or Chicago Illinois, I'm going for the creative answer of the supplier who best fits my client's budget and need and creative approach, not merely because they're close'. I was looked at as not a team player and I understand the point of those fellows.
>
> (Interviewee 16)

As this quotation suggests, some agencies have begun to look outside Detroit when forming project teams because the city's project ecology is so weak and lacking in skilled individuals. Another interviewee described the need to look beyond the city for members of a temporary project team as follows:

> Our most recent case where we sub-contracted out of the normal editing or production standpoint is we connected with a company called [company x]. They're based in New York and LA and they did all the viral work for the Blair witch project movie. We went to them and said 'you really have a great understanding of the new media world. Let's talk about ideas for Pontiac that we could collaborate on'. They were interested because we have the relationship with Pontiac which is attractive to them. We were interested because they knew this world that we didn't know and they came back with an idea where we worked with them to get [client x] on Second Life.
>
> (Interviewee 18)

Table 8.2 Employment in 2008 in Detroit Metropolitan Area (Detroit–Livonia–Dearborn) in key industries making up the advertising project ecology

Actors (NAICS 272011)	No data
Graphic design (NAICS 271024)	710
Photography (NAICS 274021)	No data
Film production (NAICS 274031)	No data
Musicians (NAICS 272042)	No data
Set designers (NAICS 271027)	No data
Post-production editing (photography, music and film) (NAICS 274032)	50
Lawyers (NAICS 231011)	3,260

Source: US Bureau of Labor Statistics May 2008 Metropolitan and Nonmetropolitan Area Occupational Employment and Wage Estimates, Macomb, Oakland and Wayne (Detroit Metropolitan Area).

Considering the discussion in the preceding chapters on New York City and Los Angeles of the importance and benefits of co-located industries and a pool of freelance temporary team members that enable the development of effective advertising, Detroit's lack of an effective project ecology is a major concern. It reinforces the apparently insulated and stagnant nature of campaigns produced in the city. Moreover, the lack of territorial assets in terms of deep pools of skilled labour and sophisticated project ecologies, as well as the absence of a large and diverse consumer audience in the city, also has implications for Detroit's network assets.

Global network assets and car campaigns

Successful advertising cities are reliant upon a range of network assets that help to generate forms of global work. It has also been noted in earlier chapters that, in many ways, the existence and nature of network assets is interrelated to the territorial assets of a city. With this in mind, at one level it might be expected that Detroit would have important network assets because of its leadership role in relation to automobile manufacturers' accounts. As such a key territorial asset, the city's client base, should generate global work. Indeed, one interviewee recalled a close cooperation with a European office for the launch of one of their car client's models:

> We developed the creative brief out of Detroit working in collaboration with the Frankfurt office. We then briefed teams both in Frankfurt and Detroit to come up with creative ideas and then [advertiser y] led the project out of here but using both our Frankfurt team and Detroit team to execute the idea that was chosen. That was really because we had clients both in Detroit here that had to approve it and in Europe it happened to be Amsterdam but we serviced the Amsterdam business out of Frankfurt.
>
> (Interview 18)

Despite the fact that Detroit agencies sometimes attract criticism from some of their US counterparts, their knowledge-base is highly valued by offices located in emerging markets such as India or China who are increasingly developing advertising campaigns for automobile manufacturers. As one interviewee suggested:

> I did work on a project for a large auto company in China; we did the work here and provided it to the office in China to review it with their client and what I was told I readily believe that in an emerging market like the People's Republic of China, there was a hunger for information on those specifics that are assumed in other developing countries.
>
> (Interviewee 16)

However, this leadership does not transform itself into the type of comprehensive network assets that New York City and Los Angeles benefit from. In the

Table 8.3 The connectivity of Detroit in terms of advertising work

Ranking of connectivity via office networks of global agencies	0.31 (New York ranks 1.0)
Score of importance of the city in terms of connectivities associated with European advertising work	–0.93 (i.e. highly insignificant)
Score of importance of the city in terms of connectivities associated with Asian advertising work	–0.47 (i.e. highly insignificant)
Number of headquarters of ten largest global agencies	0

Source: analyses presented in Chapter 4.

case of Detroit, because the city is built upon one dimensional territorial assets, relatively weak network assets exist which are also one dimensional. As Table 8.3 shows, the city falls well behind New York City and Los Angeles in terms of its network connectivity and is insignificant in terms of relational connections to either Europe or Asia. At one level this is simply because the offices of global agencies in the city almost exclusively serve automobile manufacturers, thus not creating multiple cases of account leadership spread across a range of industrial sectors. There is only so much work associated with leading campaigns for one client. At another level it is also a result of the fact that there are fewer and fewer global campaigns. Because cars are seen as a reflection, and part, of spatially entangled consumer identities – the landscape, road infrastructure and role of cars in everyday life vary tremendously between cities and regions – car adverts are tied to situated markets, something which in a post-Fordist era characterized by reflexive consumers clashes with attempts to develop universal, globally managed campaigns. Instead, transnational models have gained favour in order to respond to spatially heterogeneous consumer relationships with cars. As one interviewee noted:

> My clients are automotive clients, so their particular needs, as you might imagine, are to the size of the road, to the fuel economy to the very tangible reasons why a car is different in a place like Northern Africa or the jungles of Brazil than it is in New York or in Des Moines Iowa. So these are two mitigating factors, I guess: the need to be localized, primarily from the people who utilize the product, and the need to get economies of advertising shared voices scaled by the advertiser who's trying to sell it to places … I was recently in Berlin and there was a Mercedes taxicab. In this country there isn't a Mercedes taxicab; it is here a very prestigious premium luxury brand.
>
> (Interviewee 16)

Of course, the discussion in Chapters 6 and 7 revealed that the New York City and Los Angeles offices of global agencies also manage campaigns that require such a transnational approach and despite this still develop multidimensional network assets that generate significant amounts of global work. So the role of Detroit agencies in transnational campaigns for automobile manufacturers should not necessarily mean the city lacks network assets. Detroit lacks such assets because, first, there is not a large and diverse consumer audience in the city which producers of consumer goods wish to target and, second, the shallow labour pools and weak project ecologies mean advertising work in Detroit is often viewed as sub-standard. Hence it is rare for the Detroit offices of global agencies to be asked to work on a campaign being led by another overseas office. Put simply, advertisers in Detroit are considered to be out of touch with the rest of the USA and not the best choice when seeking a US office to collaborate and cooperate with on a transnational campaign. The result is one dimensional network assets, solely produced by and reliant upon the role of Detroit based offices in leading campaigns for the 'Big Three'.

Somewhat worryingly, Detroit's control over worldwide campaigns for their automobile manufacturer clients is increasingly coming under threat. As two interviewees put it:

> It's sort of going that way right now as the car industry falls into more trouble, our influence in Detroit is becoming much more limited. We used to have the largest [agency x] office, even larger than New York, because our number one client's right across the street. But then they began dividing up the company and we have less influence. Chicago, LA and New York are far ahead of us now.
>
> (Interviewee 17)

> It's interesting to see the ebbs and flows in global advertising. Depending on who's in charge, somebody will come in and say 'I want it to be centralised and have everything come out of New York or Detroit or wherever' and it will occur that way for three or four years and then the next person that comes in charge says 'no I'll put all the budgets out there in the local markets and it's all gonna happen to be there'.
>
> (Interviewee 18)

The transitions described by these two advertisers are in train as we write and can be explained in two main ways. First and further driving Detroit's declining role as a leader of global campaigns for automobile industry clients, is the fact that transnational campaigns increasingly involve regionalization. The 'Big Three' have responded to the place-specific entanglements that affect consumers' relationships with cars by splitting campaigns into regional blocks with Detroit increasingly being responsible only for campaigns in the Americas or even North America only. There are exceptions to this: Ford for example continues to develop the global car concept, but predominantly campaigns and products have regional geographies (on this see Dicken 2007). So as one advertiser lamented:

> This was a choice of the client. We had nothing to do with it. We lost the [brand x] business so at that point they made the decision to not try to run it as a global piece of business out of Detroit but that they would just let the individual markets handle it ... with [client x], ultimately the decisions are made on the local ground level, whether it be in Zurich or Shanghai. That's where the decisions are being made, the money is being spent, the resources deployed.
>
> (Interviewee 18)

Second, agencies in Detroit are increasingly viewed by automobile clients as a sub-optimal choice because of the effects of shallow labour pools and unsophisticated project ecologies on the work produced. Client accounts for brands such as Jaguar, Aston Martin and Land Rover – all brands at various points in time owned by one of the 'Big Three' – have been transferred to Los Angeles because of the importance of the advertising labour pools and project ecology in the city and the knowledge these provide of Californian consumers. Other US cities have also benefited at the expense of Detroit, again because of their superior territorial assets. As two interviewees noted:

> We had the [brand x] business for 75 years and we lost it to [agency x] in Boston. We had 200 people on the business and they were a total of about 20 people. We lost that business to them.
>
> (Interviewee 18)

> We just lost business to an agency called [agency y]. [Agency y] is 30 people in San Francisco ... We've had the [client y] account for 20 years and they just kind of surprisingly said 'we wanna take a fresh look at the [product y] brand and we're gonna ask you and three other agencies to pitch it'. This agency in San Francisco won the pitch so that was very new for us and it's very recent too. They haven't even produced a campaign yet, it's so new. That was obviously a big change for us because we'd never lost something like that.
>
> (Interviewee 20)

Reinforcing this problem is another one of Detroit's frailties in terms of network assets. The stagnation within the city's labour pools and the weakness of the project ecology have become self reproducing as advertisers from within and without the USA are put off moving to the city because of its poor reputation in advertising circles. As one advertiser put it, 'it's not a glamorous city. It has one industry that isn't cutting edge and generally it's not accepted that automobiles advertising set trends in the advertising business. It's an industry that requires a lot of support' (Interviewee 16). A key network asset of a successful advertising city is labour flows into the city, generated by agencies and the reputation of the city itself. As Detroit lacks such an asset, it is very difficult to inject new energy and ideas into agencies or to develop the city's project ecology. One interviewee described this challenge in the following terms:

> In Detroit, you'll find that it's pretty hard to find a diversity of non US nationals because if somebody is moving from Paris or London, they're not gonna choose Detroit to go to. They're gonna go to New York. When I worked at the New York office for three years, I felt we had pretty big diversity there for non-US people … Detroit is a tough place to recruit people to because youth wants to go to New York or LA, the glamorous places. You have to find somebody like me who has a passion for cars or has ties to Detroit or you just find interesting things you're able to offer.
>
> (Interviewee 18)

Attempts have been made to overcome the limitations in Detroit's territorial and network assets and to create the conditions for the city's recovery, notably with the help of the Time Inc. unit of Time Warner which launched a year-long 'Assignment Detroit' project involving reporters from various magazines such as CNN Money (see CNN 2010). The related initiative 'Selling Detroit' launched in November 2009, at a cost of US$400,000, and aimed to attract young, entrepreneurial and creative talent in fields such as design, architecture, music, fashion, the arts and of course advertising, to live and work in Detroit. Five of the main advertising agencies in the city agreed to take part in the contest, with campaigns ironically this time focused on rebranding their own city's image. The contest was won by McCann Erickson's campaign 'Creativity lives in Detroit'. The print ad which appeared in *Fortune* magazine as well as on several specialized websites spells out the word 'Detroit' with a mosaic of examples of local achievements, from Pewabic Pottery, founded in 1903, the Detroit Derby Girls, an all-female, amateur, flat-track roller derby league, to the Von Bondies, a local rock band (see Figure 8.2). The mosaic is full of hyperlinks to an accompanying website to be launched in the course of 2010. The text of the advert, posted online at www.creativitylivesindetroit.com (accessed 1 September 2010), reads:

> This city is built on creativity. You can hear it in our music. You can see it in our arts. You can taste it in our restaurants. And you can feel it in our streets with creative people just like you. Come build something special.

Whether this new initiative will suffice to turn around the fate of the Nation's Motor City remains open for debate. As one interviewee argued:

> There's been a kind of an underground movement in Detroit, a kind of guerilla movement to re-establish it as a centre of creativity amongst all this gloom and disaster. I'm not sure how successful that's being, it's relatively new initiative. Some of the small agencies in town are trying to get this ground swell movement together saying: 'Detroit is a gritty place, it's full of real people, they are creative, so let's recreate the city'. But there's some of the other forces working against them. It's hard to see how successful that will be. Every time you switch on the news, there's some kind of bad news

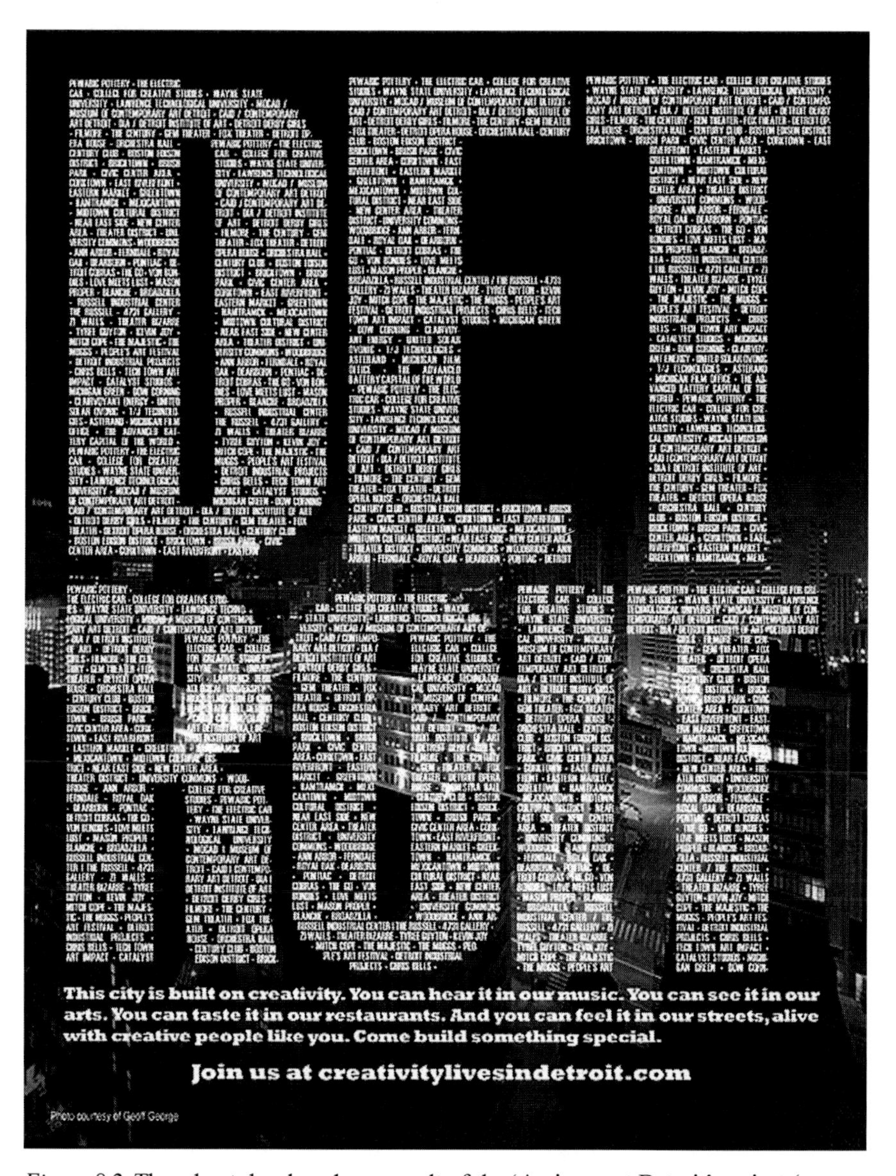

Figure 8.2 The advert developed as a result of the 'Assignment Detroit' project (source: www.creativitylivesindetroit.com (accessed 1 September 2010)).

about Detroit, whether it's the automotive industry or foreclosures or people trying to explode bombs in aeroplanes over Detroit. So we're still a bit of a figure of fun a cynicism from a lot of other people.

(Interviewee 30)

Implications

Detroit as an advertising city looks increasingly fragile. It is surprising to be talking in such terms about a city that, in 2000, was ranked as the sixth largest city in the world for advertising billing, above cities such as Milan, Frankfurt and Sydney. But this reveals how misleading analyses that purely focus upon territorial assets can be. Measures of success that do not take account of the wider network processes of city formation often hide important weaknesses. The discussion here suggests that, together, a combination of one dimensional territorial assets that lead to one dimensional network assets is responsible for Detroit's problems. Specifically Detroit's position as a leading advertising city is overly reliant on a historical legacy associated with particular geographical conditions – the agglomeration of auto manufacturers in the city and their use of co-located agencies to develop global command and control campaigns. These conditions have begun to change and hence the city's future role in advertising globalization looks shaky. As such, Detroit and its stagnation is the reverse of New York City which has developed a new role in advertising globalization in the twenty-first century. And as a result, Detroit has begun to experience globalization as a process through which work is lost as a result of 'competition' from other cities. Detroit has found itself in a position whereby processes of spatial rearrangement

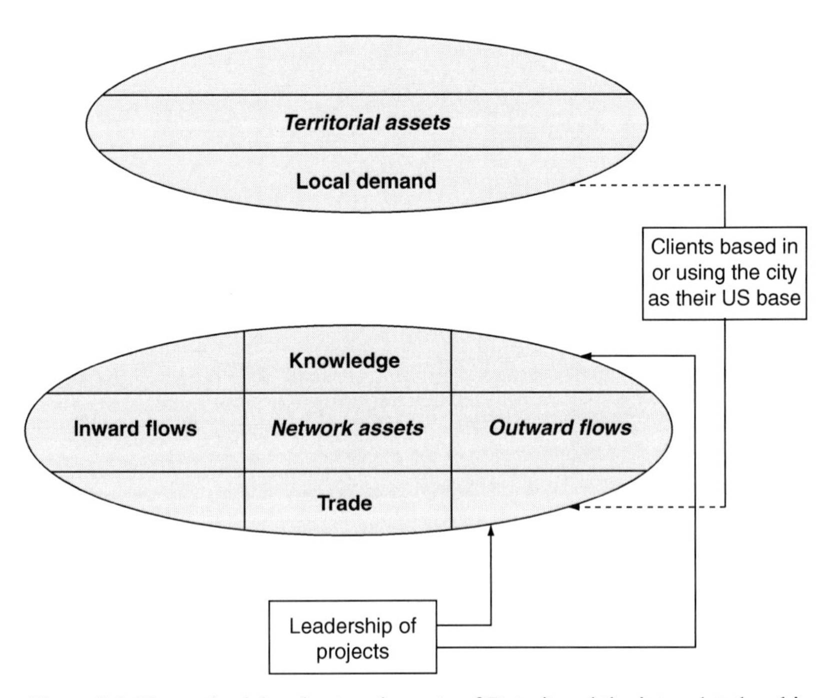

Figure 8.3 The territorial and network assets of Detroit and the interrelated and inseparable ways that they define the city's role in the geographies of advertising globalization.

undermine local work which is not replaced by new global work. This reveals why networked cities are successful cities in the twenty-first century: work generated through relational connections is as, if not more, important than locally generated work, and is key to constructing a city with a successful role in the space economy.

The example of Detroit further reveals, then, the usefulness of our theoretical approach which explores the way an analysis of territorial and network assets can be used to explain the failings and weakness of a city and its role in advertising globalization. Figure 8.3 captures the explanation offered by the theoretical framework for Detroit's role in advertising globalization and the challenges it faces. It is clear that Detroit's territorial and network assets are much less complex, lack the multidimensional nature and feedbacks that exist in the case of New York City (Figure 6.2) and Los Angeles (Figure 7.2), and thus result in the city being reliant on the continued existence of a particular set of economic foundations. This seems to suggest that Detroit's advertising economy is somewhat fragile. If one of the components of Figure 8.3 was removed, little would remain in terms of the city's advertising economy.

Conclusions

Detroit appears to provide an example of how territorialized demand on its own – i.e. a powerful local client base – cannot act as the basis for the formation of a city that plays a major role in advertising globalization. In addition, networked supply – leading to additional demand and deep pools of knowledgeable workers and sophisticated project ecologies – is also needed. This has implications for debates about the role of the creative class (Florida 2002), the relationship between the creative class and demand (Pratt 2008), and wider debates about connections between related and unrelated variety, spillovers and city development (see for example, Boschma 2005; Glaeser *et al.* 1992; Jacobs 1969, 1984). All of these issues are discussed further in the final chapter of the book.

In the next chapter we revisit Detroit, along with New York City and Los Angeles, to consider how the credit crisis and recession have affected the processes discussed in this chapter and Chapters 6 and 7. As the discussion here of Detroit's challenges suggests, of particular significance are questions about the way the territorial and network assets of a city are affected by and determine the effects of recession and economic crisis.

9 Coda

Agencies, cities and recession

The research reported in this book was conducted in 2007 and prior to the credit crisis and ensuing global recession that has had major impacts on all sectors of the world economy. As a result, we are not in a position to provide a detailed synopsis of how the recession has impacted upon the advertising industry, not least because at the time of writing in early 2010 agencies were still in the process of adapting to the changing strategies and demands of clients in this new economic world. However, we are able to paint a picture of the way the recession has affected our studied global agencies and three case study cities, New York, Los Angeles and Detroit, drawn from a number of follow-up interviews with advertisers originally spoken to during data collection in 2007. These insights are supplemented by analysis of media reports of changes in advertising spending and agency strategies. In addition, our interpretations are enhanced through being able to draw on the argument developed in earlier chapters about the way territorial and network assets define the role of a city in advertising globalization to explain the differing impacts of the recession on advertising work in New York, Los Angeles and Detroit. In short, the recession helps to further confirm that the cities with the strongest and most resilient advertising economies are those with robust territorial and network assets.

In this coda we, therefore, consider the impacts of the recession on the advertising industry, with particular focus on the US context, and then draw upon our follow-up interviews to analyse the specific impacts on New York, Los Angeles and Detroit during 2008 and 2009.

Advertising and the global recession

This is not the place to review the generic economic effects of the recession on the world economy. Suffice to say that a 12 per cent slump in global trade (see UNCTAD 2009), the fastest decline since the Second World War, and a 20 per cent decline in levels of foreign direct investment in 2008 when compared with 2007, a decline expected to be worse in 2009 but at the time of writing unreported, have had significant effects on the advertising industry in the USA and worldwide. As have related steep declines in consumer spending growth of over 1 per cent in 2009 in the USA. Together such trends led to many of the clients of global agencies reducing advertising budgets as a way of maintaining profits or minimizing losses despite advertising executives' protests that clients should actually maintain spending in order to ensure more selective consumers choose their products.

The available figures that capture the direct effects of the recession on the advertising industry reveal major upheavals. In 2009, all of the major holding companies reported declines in revenues (Table 9.1) with all regions experiencing declines, including Asia: even the Chinese economy slowed although with a much less significant impact than the slowdown in the USA and Western Europe. As Table 9.1 also shows, in the US context the recession meant major job losses in the advertising industry with levels of employment falling by almost 10 per cent between 2006 and 2010 and to the lowest levels seen in the past two decades. As a result, it was widely acknowledged by advertising executives, both in the media and in the follow-up interviews we completed, that the recession resulting from the credit crisis was the worst ever experienced by the advertising industry.

With regards to the focus of the book, namely the role of cities in advertising globalization, one of the most important features of the recession has been its uneven impact. At the time of writing, city-level data on change in employment in advertising was not available. But again, through reports in the media and our follow-up interviews, it is possible to reveal a picture of city-specific degrees of decline and upheaval. In the case of the USA, explanation of intra-national variations in the impacts of the slowdown on advertising cities is partly provided by Table 9.1. Whilst clients from all sectors reduced advertising spending in 2009, some clients cut budgets more radically than others. The automotive industry is particularly significant in this regard. Whilst not showing the largest percentage decline, the impact of cuts in this sector have been significant because of the sheer size of advertising budgets in the automotive industry, totalling US$15.6 billion in 2009 (*Advertising Age* 2010c). Thus large amounts of revenue were lost by agencies as a result of the cuts imposed by firms such as Ford and Toyota.

Table 9.1 The impacts of the recession on the advertising industry

Change in holding group revenues (2008–09)[1]	*Sectors experiencing most severe declines in advertising spending 2007–08*[2]	*Changes in employment in advertising agencies (NAICS 54181) in the USA*[3]
WPP: –8% Dentsu: –44% Havas: –8% Interpublic: –13% Omnicom: –15% Publicis: –6.5%	Automotive: –15% Real estate: –30% Office equipment: –30% Home and building supplies: –16% Retail: –6%	2008–09: –9,600 2009–10: –19,900

Notes:
1 Financial reports available from groups' websites.
2 *Advertising Age* 2010c.
3 *Advertising Age* 2010b.

Offices of global agencies that were particularly reliant on spending by automotive clients, such as Detroit, were, therefore, particularly hard hit by the recession. But, it is not only the different client bases of cities that have caused variations in the impacts of the recession on cities.

Variations in other territorial and network assets of cities beyond client base have also been pivotal in determining how resilient advertising industries have been in different cities. In order to understand how a city's territorial and network assets have influenced the effects of the recession on advertising work, we therefore consider the main changes in the organization of global advertising work as a result of the recession and the effects on our three case study cities: New York, Los Angeles and Detroit.

Changes in the organization of advertising work

Perhaps one of the most significant developments in the organization of advertising work as a result of the recession relates to the use of transnational, collaborative and cooperative campaign structures. As discussed throughout the book, the early part of the twenty-first century saw agencies move away from global, imperial models of organizing work in response to the challenge of targeting reflexive, spatially entangled consumers. However, as also acknowledged in discussions, transnational campaigns are expensive and complicated to organize. As a result, before the recession and in response to both fierce domestic and international competition, global agencies were already talking about the importance of being 'lean' in their functioning and the implications of this for consolidation, and the continuity of the transnational integration between offices that had grown in the late twentieth and early twentieth centuries. Such debates are aptly illustrated by one follow-up interviewee who suggested that:

> In so far as we are concerned … I think around the world you are going to get some consolidation because I think as opposed to each country having its own agency you can actually combine. In some regions you can develop a hybrid agency of a number of different nationals who could perhaps work on some assignments. I think you will get centres of excellence like that. I think [agency x] is doing that. [Agency x] has cut down on the number of offices they have. You used to have an office wherever there was a client. I think that will change, the economics forced that change. The way we can all communicate now makes it much easier. Well it's going two ways … we have separate operating companies and the other functions. Other people have put them all together in one building and one office. Personally I think that's a better way to go and I think it will perhaps go that way. Those are the two models that are competing.
>
> (Interviewee 24)

The recession and the need to deliver 'more for less' to clients has seen these debates grow more heated and one result has been that the global command and control model has been rejuvenated. This does not mean transnational campaigns have all been replaced by global campaigns. Rather it means that in 2008 and 2009 the global model, which was out of fashion in the preceding years, has become more fashionable after a period of decline. Centralization of campaigns in one office per region has been used as a way of cutting costs, most notably because it allows headcounts to be reduced in many offices. In effect, agencies have reverted to the basic logic of economies of scale. As one advertiser described in a follow up interview: 'There is room for integration and centralization. If you're the office that's got the centralized account, in a way, you might benefit. Certainly from my office's point of view, it's [campaign management] become much more national' (Interviewee 30). As this quotation suggests, at the level of campaigns, the effects of reverting to command and control management are felt both intra- and inter-nationally. In the US context, nationwide campaigns that may have been managed by multiple offices before the recession in order to respond to regionally heterogeneous consumers are increasingly managed by one office. Similarly, there is a greater reliance on pan-regional campaigns in arenas such as Europe and South America. As one interviewee described:

> You absolutely have [to] flatten the world at a time like this and that's where having a high quality network with lots of decent agencies and lots of very good people becomes a real asset because you can take advantage of both capacity and in some cases lower costs in different markets. We've been doing more of that. We always intended to but we've probably done it better and faster in the last 15 months as a consequence of the recession.
>
> (Interviewee 5)

As this quotation suggests, as agencies have sought to centralize work, it is the office/city that has the best assets which can be leveraged on campaigns using a global (pan-regional) structure that have benefited from restructured work processes. Lead offices have, therefore, maintained their important role and selected one office in a region, or multiple offices, but fewer than in the past when transnational campaigns were more common, because of their ability to provide, for example, outstanding creative work for a pan-regional campaign or, in some cases, cheaper creative work than other offices in the same region.

In addition, there has been a second key change in the organization of advertising work as a result of the recession. As knowledge-intensive business service firms, the main cost of advertising agencies is the workforce. Consequently strategic management, planning and creative workers have been the target of cost cutting as agencies have sought to reduce headcounts in most offices. For example, the WPP group of companies cut over 14,000 jobs in 2009 (WPP 2010). Such reductions in headcount are, at one level, a result of the combined effects of reduced demand and revenues and the rejuvenation of the global model of campaign management. However, at another level cuts are also a result of agencies externalizing more services and exploiting urban project ecologies. Agencies have turned to the use of temporary teams, as described in Chapter 3, with freelancers and independent firms being called on more, and competing, in particular, in planning and creative tasks that would previously have been handled by employees of an agency. The benefit for agencies of exploiting urban project ecologies is that individuals can be called on as and when needed, thus reducing the overheads the agency carries on a permanent basis. In a follow-up interview one account planner argued that:

> We've made more use of freelance people than before. Certainly the freelance market in New York has been quite buoyant in 2009 and people have found it possible to get work that way ... in certain businesses there were hiring freezes and just across the board, they're mandatory, you can hire anyone. I've heard every kind of cost limitation strategy being employed including as you say not hiring people full-time but bringing them in freelance.
>
> (Interviewee 6)

Of course, for individual workers the adoption of such a strategy can be highly problematic. Workers once employed by an agency on a full time basis have found themselves unemployed and part of a growing pool of freelancers constantly facing the threat of periods without pay.

In terms of the impacts on cities, the greater use of urban project ecologies by agencies has benefited those cities with strong ecologies comprising a wide range of skilled workers, thus ensuring certain cities' continued centrality in the geographies of advertising globalization. Of course, this means those cities without such deep and rich project ecologies become excluded for advertising work. Next we consider the implications of these two changes in the organization of advertising work, as well as a broader array of geographically contingent variables, on the experience of the recession in New York, Los Angeles and Detroit.

Changes in city roles

In order to understand the fates of our case study cities it is important to return to the preceding discussion about the changing organization of advertising work and the book's meta-narrative about the territorial and network assets of cities. The recession has proven an ideal testing grounding for the thesis laid out in Chapter 5 that a combination of territorial and network assets determine a city's role in advertising globalization. Specifically, the resilience of a city in times of recession, as well its strength in times of economic stability or boom, has been shown to relate to the strength of a city's assets. Cities possessing both strong territorial and network assets are more resilient whilst those with weaker assets have had to adapt or have been particularly vulnerable to the effects of the recession. Table 9.2 captures this dynamic by considering the different territorial and network assets, discussed in Chapter 5, and their role in explaining why our three case study cities, with differential assets, have experienced the recession in distinctive

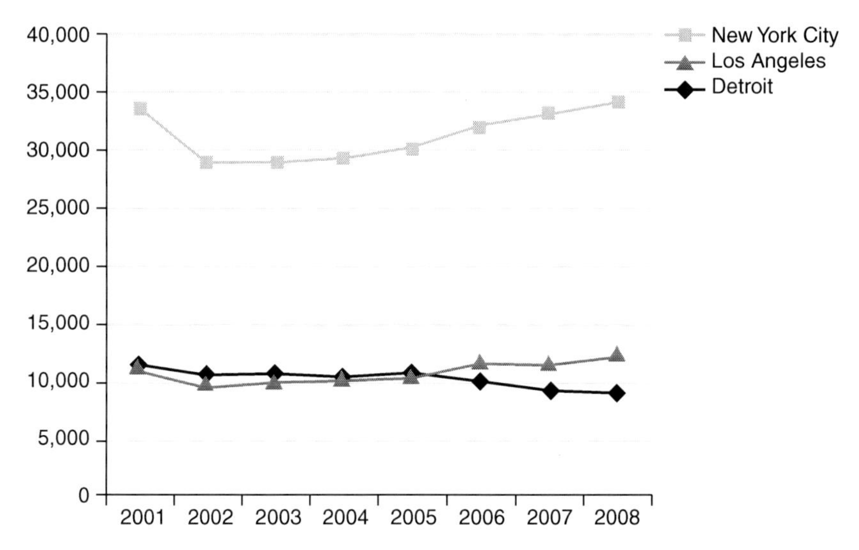

Figure 9.1 Employment in advertising agencies in New York City, Los Angeles and Detroit, 2001–08 (source: US Bureau of Labor Statistics Metropolitan and Nonmetropolitan Area Occupational Employment and Wage Estimates (NAICS 54181: 'Advertising Agencies' including activities such as public relations, media buying, display advertising, direct mail and advertising distribution)).

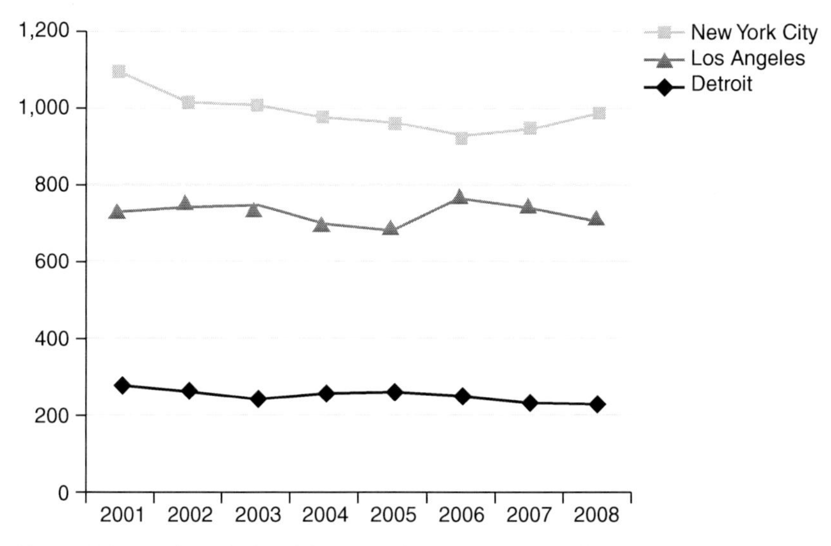

Figure 9.2 Number of advertising agencies in New York City, Los Angeles and Detroit, 2001–08 (source: US Bureau of Labor Statistics Metropolitan and Nonmetropolitan Area Occupational Employment and Wage Estimates (NAICS 54181: 'Advertising Agencies' including activities such as public relations, media buying, display advertising, direct mail and advertising distribution)).

Table 9.2 The effects of the recession, as determined by their possession of the territorial and network assets that define a key advertising city

	Territorial assets		*Network assets*	*Case study US city*	*Change in network connectivity ranking 2000–08*
Resilient city	*Client base*: large and diverse ensuring not reliant on one type of product which may or may not be affected severely *Pools of skilled labour*: deep and part of sophisticated project ecology that continues to make city attractive to clients and agencies *Consumer markets*: large and heterogeneous ensuring continued importance of the office's city/region in marketing strategies	→ →	*Leadership in global campaigns*: continued leadership *Membership of global campaign teams*: key pan-regional node in global campaigns led by other offices; collaborator on remaining transnational campaigns	New York	0
Adapting city	*Client base*: smaller resulting in changes in demand from one client having bigger impacts but diverse enough to prevent fatal declines in demand *Pools of skilled labour*: sufficiently deep to allow clients' needs to be met *Consumer markets*: large and heterogeneous ensuring continued importance of the office's city/region in marketing strategies	→ →	*Leadership in global campaigns*: continued role but less significant than resilient city due to smaller client base which means less likelihood of significant amounts of global work *Membership of global campaign teams*: continued but diminishing role as collaborator on transnational campaigns due to rejuvenation of global management strategy and out competition by resilient cities	Los Angeles	–0.07
Vulnerable city	*Client base*: lacking in diversity meaning effects of downturn may be exaggerated *Pools of skilled labour*: limited and continuously shrinking *Consumer markets*: small and unsophisticated	→ →	*Leadership in global campaigns*: limited and reliant on demand for global advertising from one dimensional client base *Membership of global campaign teams*: little due to shallow labour pools, unsophisticated project ecologies and unimportant consumer markets	Detroit	–0.13

ways. Meanwhile, Figures 9.1 and 9.2 show changing levels of advertising employment and agency numbers in our three case study cities, between 2001 and 2008. This data shows the effects of the 2001 dot.com bust and associated mini recession on the advertising industries in the three cities, but the beginnings of the effects of the credit crisis and ensuing recession of 2007 onwards are not, as yet, visible. At the time of writing city level data for employment in 2009 was not available. We, therefore, rely on our follow-up interviews to make sense of trends in 2009 in the three cities.

New York

New York has shown itself to be a relatively resilient city during the recession. This does not mean the city has not seen the workforce in agencies shrink and office revenues decline. Rather, it means that it has been able to maintain its role as a pre-eminent advertising city without major upheavals in the structure of agencies or the associated urban project ecology. New York's territorial assets ensure continued demand for advertising *managed* by workers in the city. This relates to the city's critical mass of clients with a diverse array of products. Whilst some clients have drastically cut budgets, the diversity of the client base means major cuts by one client, whilst leading to some job losses, are less problematic if offset by sustained levels of spending by other clients who continue to turn to the New York offices of agencies for campaigns. As one interviewee affirmed when discussing the recession:

> New York's done better [than other cities during the recession], there's no denying it. There are a number of factors at play. Some clients have consolidated businesses with the office instead of spreading it across a number of offices ... A lot of our business here in NY for example is with companies like [major global client x] or [major global client y] who sell staples, the products we all need. They're not really recession sensitive. If you need toothpaste, you need toothpaste.
>
> (Interviewee 6)

As the quotation suggests, New York's territorial assets also generate network assets in terms of a continued if slightly diminished role as the lead office for the management of worldwide campaigns, thus ensuring agencies in the city suffer less from rationalizations made as part of an attempt to involve fewer offices in campaigns as global management strategies are rejuvenated. Relatedly, in terms of the delivery of campaigns, because New York has territorial assets in the shape of such a large consumer base, in the city and throughout the greater east coast region, with deep pools of labour and a sophisticated project ecology, agencies' offices in the city also act as nodes in worldwide campaigns led by other offices. Specifically the territorial assets ensure New York agencies are often chosen as *the* key US node in globally managed campaigns as well as one amongst many US nodes in transnational campaigns. The territorial and network assets of the city combine, then, to help make the city resilient, although not immune, during recession.

Los Angeles

Exemplifying a subtly different experience of the recession, Los Angeles has had to adapt and engage in more significant restructuring than New York. Adaptation has been necessary for multiple reasons. In terms of territorial assets, Los Angeles possesses a *relatively* less sophisticated client base. Nonetheless, whilst not having the strength in depth of New York, the client base has been diverse enough to sustain respectable revenues for the offices of agencies. As one interviewee commented in a follow-up discussion about the recession, 'We won a very large power company. It's based in Southern California. It's called [client x]. It's spends quite a lot of money. Of course, they've not been affected by the downturn of the economy too much' (Interviewee 30). However, other major clients, such as some of the Japanese car manufacturers who use California and the Los Angeles offices of agencies as the basing point for US marketing work, have been heavily hit and reductions in revenues from these clients has impacted on agencies significantly, requiring restructuring and redundancies involving more significant upheavals than in the case of New York.

More significantly, though, has been the way Los Angeles' network assets have determined the impacts of the recession on advertising work. At one level, Los Angeles has maintained its role as the lead office for campaigns for a number of major clients. But, the rejuvenated global strategy has often led to a reduced role for Los Angeles as a collaborator on worldwide campaigns. In particular, the reduction in the number of adverts tailored to west coast consumers and a tendency to choose New York as the US node for global campaigns, has dented the revenues of the Los Angeles offices of agencies. For Los Angeles, adaptation is needed because of the way territorial and network assets have been modified during the recession to generate a subtly different role for the city in advertising globalization.

Detroit

In contrast to resilient and adapting cities, Detroit has experienced the recession very differently as a vulnerable city. By 2008, Metropolitan Detroit had already lost a significant number of advertising jobs compared with earlier in the century (see Chapter 8). Data for 2009 when released is likely to show even further declines because at the heart of Detroit's vulnerability to the recession is its fragile territorial assets and their impacts on the city's network assets. As described in Chapter 8, Detroit's advertising industry has been almost exclusively reliant on an automobile industry client base for many years and the decline in jobs in advertising over recent years results almost exclusively from this reliance. The financial crisis, intense competition from overseas producers, the restructuring of supply chains and subsequent locational adjustments, have recently led to the decline of the traditional domestic car makers (General Motors and Chrysler in particular). The one dimensional client related territorial asset of the city has meant the advertising industry has been susceptible to reductions in advertising budgets in the automobile industry, an industry which in 2009 changed from being a relatively prosperous industry to an industry struggling to stay afloat. Bankruptcies and bailouts have plagued US automakers and the Big Three implemented major changes in their advertising strategies, principally meaning reductions in the number of campaigns and major cuts in budgets for the campaigns that remained (Table 9.2). With no other clients to cushion the shock of this blow, agencies in Detroit have become vulnerable to downsizing and closure.

Most worryingly, as part of the rejuvenation of global campaign management strategies, during the recession Detroit has even begun to lose its role as the leader of automobile industry campaigns and hence has seen the destruction of the weak network assets that existed. Some brands, such as GM's Saab division, vanished altogether. Saab's agency was McCann in Detroit. More concerning, though, is the acceleration in the process of auto manufacturers ditching Detroit as the base for their US marketing. The Chrysler Group is certainly the one that has shown the most wanderlust for agencies outside Detroit. From the beginning of 2010, advertising for the Chrysler brand is being led by the Minneapolis based agency Fallon, while the Dodge, Jeep and Ram truck accounts are being headed by a trio of US independent firms: Portland based Wieden & Kennedy, GlobalHue (the only Detroit agency continuing to work for Chrysler) and Dallas based Richards Group, respectively. Again reflecting the ideas outlined in Chapter 8 about Detroit's shortcomings in terms of talent and creativity, auto industry clients have decided that US-wide campaigns are best led by offices based outside Detroit, although, interestingly, not offices of global agencies in Los Angeles or New York. According to Olivier François, Chrysler's head of marketing, this trend can be explained by the fact that, 'the brands retain individual agencies that fit with the brand direction. Individual agencies also will enable us to deliver unique ideas and broaden our reach to existing and potential customers' (*Wall Street Journal* 2010). Figure 9.3 details this mass migration of work from Detroit.

Confirming the importance of network assets generated by leadership of worldwide campaigns, the most prosperous agency in Detroit at the time of writing, WPP's 'Team Detroit' made up on multiple agencies, relies on its leadership role in Ford's global campaigns. As one account planner noted:

> They're launching a new version of the Ford Focus in 2011 which will essentially be the same vehicle everywhere. It might look a little bit different, a slightly different trim, but essentially it's the same vehicle everywhere. WPP and Team Detroit is part of the global launch of the vehicle ... So everybody is increasingly straddling the local market and the global market to survive.
>
> (Interviewee 15)

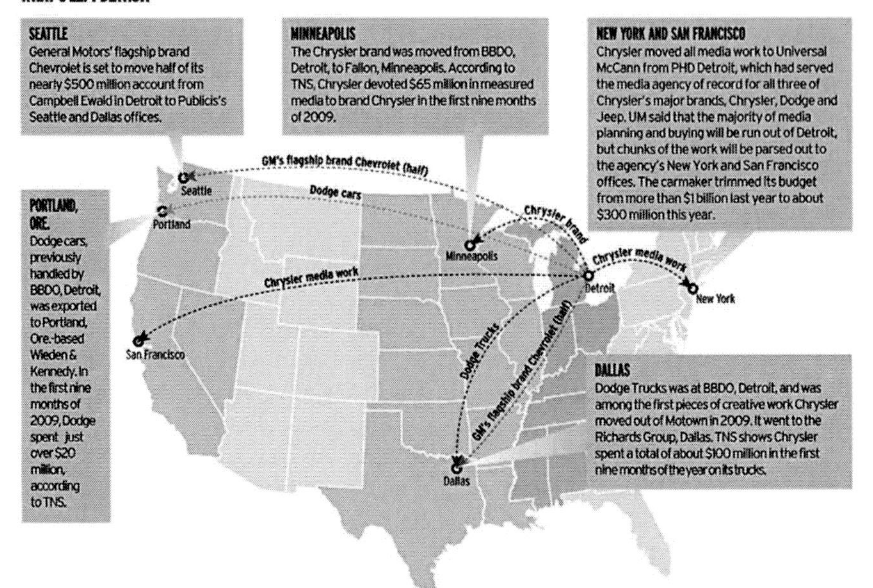

Figure 9.3 Advertising accounts of the major auto manufacturers that left Detroit in 2009/10 (source: *Advertising Age*).

The Ford Motor Co., which distributes its three main brands (Ford, Lincoln and Mercury) across six continents with a workforce of 200,000 employees and about ninety plants worldwide is, however, unique for a couple of major reasons. The first is its apparent success at designing cars which fit the global market as opposed to solely appealing to the Midwestern consumer taste. The second is that Ford, unlike GM and Chrysler, did not rely on President Obama's rescue plan for the car industry. In the words of one interviewee:

> From a Ford global perspective, there's lots of relationships with the Ford business in India and the Ford business in China. Increasingly the vehicles are more global. It doesn't mean they all look the same but underneath the mechanicals and the components are increasingly global so the vehicle might just get a slightly different skin on it in one market versus another. … [Ford] got a lot of credit for not taking bail-out money from the government and for having a plan and making it work and having some good vehicles coming in the next year or so. Ford is on a bit of a roll at the moment fortunately. We can be part of that so we're doing better than the other Detroit agencies at the moment.
>
> (Interview 15)

Nonetheless, the business of Ford alone cannot sustain Detroit or overcome all of its problems.

Further exaggerating Detroit's vulnerability to the recession is the fact that, as highlighted in Chapter 8, because the city has never been associated with outstanding advertising talent it has never developed network assets in the shape of a role as a collaborator on worldwide campaigns. During the recession this has reinforced the city's vulnerability as the cushion that can be provided by work on campaigns led by other offices has also been missing. Detroit has been too focused in its advertising work, capable of serving the needs of 'local' motor industry clients but lacking diversified capabilities. As two interviewees put it:

> the old established agency dinosaurs are largely breaking up here and people are looking to give their work out to more places rather than consolidating in one place because they believe they get better ideas from more people. And then there's a lot of the support network that's struggling as well. A lot of the digital work is increasingly being outsourced to South America or the Far East.
>
> (Interviewee 15)

> Within advertising I wouldn't say that one particular city has been hit badly other than the obvious city of Detroit. It would be a case by case basis. It depends on what accounts you have. If you have accounts that have been hit hard like in automobile, it's terrible. If you are in wireless like Verizon or ICT or pharmaceuticals, not so bad. It depends on the market you're in.
>
> (Interviewee 30)

A total of 35 per cent of Detroit's municipal area was uninhabited in 2009 and it is the only city in the world to have lost more than half its population – almost a million people – in half a century. Only around the university can a few pedestrians be seen, or at the end of the school day, on the main avenues of Woodward, Michigan or Gratiot. The sub-prime crisis has increased the problem. Michigan's largest city was one of the worst affected by the sale of sub-prime mortgages. The failure of thousands of borrowers to keep up the rising monthly payments led to repossessions – 67,000 between 2005 and 2008, according to the city council (Popelard and Vannier 2010). Capitalism's latest crisis affected, then, the people of Detroit particularly badly. Residents were hit both by the financial meltdown and the fall in car production that followed as the collapse of the banking system reduced access to credit, consumption decreased causing a dramatic fall in car sales in the US and the 'Big Three' were hit with falling sales. Over-indebted, under-capitalized and out-competed by Japanese manufacturers, GM, Chrysler and Ford relied for their survival upon the US government's federal rescue plan. But the plan didn't prevent job losses; in fact it insisted on them to make the companies viable for the future. Unemployment almost doubled in Detroit between January 2008 and July 2009, rising from 14.8 per cent to 28.9 per cent. Yet, the real rate may be over 40 per cent due to under reporting (Popelard and Vannier 2010).

Table 9.3 Main agencies leaving Detroit in 2009/10

Agency	*Group*	*Clients/accounts*	*Billings (first three quarters of 2009)*
WPP Team Detroit (JWT, Ogilvy, Y&R, Mindshare, Wunderman)	WPP	Ford, Lincoln and Mercury strategy, creative and media	$729 million
Campbell-Ewald	Omnicom	Lost Chevy cars to Publicis Keeps Chevrolet truck advertising for the Olympics Assignment to create Olympic ads for Eco fuel-efficiency vehicles	$250 million
Leo Burnett	Publicis	GM's Buick and GMC national creative accounts	$200 million
McCann Erickson	Omnicom	GM corporate work	$170 million
MRM	Interpublic	GM's digital ad work Account to be staffed by hiring 200 people between Detroit and Argentina	$100 million in billings over five years

Source: Advertising Age (2010a).

In terms of advertising, the worst case scenario for the city has come true: agencies have closed their doors. An estimated $500 million to $1 billion in advertising dollars were sucked out of Detroit agencies in the 2007 and 2009 period. As a result, thousands of advertising professionals lost their jobs (*Advertising Age* 2010a). BBDO was just one of the agencies which closed its doors after servicing Chrysler in the city for many years. According to our interviews, the agency saw staff levels fall from 2000 in the year 2000, to 900 in 2007 and down to 500 at the time of the closure. Table 9.3 provides details of the agencies that were left in Detroit at the time of writing with many in the industry wondering how long they will remain in the city. In sum:

> Lots of people are looking at Detroit as a bit of a dinosaur in many ways these days. They're starting to say 'I don't need just because I have an automotive account, I don't need to do that in Detroit anymore, I can do that anywhere'. And they are smarter people in other places they think.
>
> (Interviewee 15)

Conclusions

In this coda chapter we have presented what is inevitably a partial and limited story about the effects of the credit crisis and recession on the globalization of advertising and on our three case study cities. Based on limited data we have sought to capture some of the most significant changes in the organization of advertising work and the way these changes have interacted with the territorial and network assets of New York, Los Angeles and Detroit to generate city-specific experiences of the recession. As such, the framework we have developed in the book and applied here to explore the effects of the recession on the three cities could be applied to other cities worldwide to understand their resilience, adaptation or vulnerability to the recession. There will, of course, be other city-specific contingencies that also need to be factored into the analysis. But the basic principles of exploring territorial and network assets should hold.

With regard to characterizations of the effects of the recession, one of our follow up interviewees offered the following framing:

> to some extent this has been an asymmetric recession, the impact of it on agencies has been asymmetric as well as cities. In the extreme case of Detroit, we're closing it. In other places, we're growing. There are no straight lines and there are no averages.
>
> (Interviewee 5)

The asymmetries described by this interviewee can be seen in the resilience, adaptation and vulnerability to the recession of New York, Los Angeles and Detroit respectively. Each city's territorial and network assets together determine the way economic change impacts on the role of a city in processes of advertising globalization and the outcomes in terms of advertising work in the city. In particular the asymmetries suggest that, in contrast to periods of stability or expansion when new advertising work is being generated and cities cooperate rather than compete as part of transnational models, in the recent period of recession changes in the organization of advertising globalization have moved towards a zero sum game.

Conclusions

Advertising agencies and cities in the space economy

Through the theoretically informed rich and innovative empirical analysis presented in this book, we have sought to unpack the contemporary geographies of advertising globalization. In focusing on three US cities, New York, Los Angeles and Detroit, our intention has not been to produce an analysis of advertising globalization solely relevant to the US context. Rather, we have used our three case study cities to exemplify *processes* of advertising globalization that have worldwide relevance, and that can be used to explain the way agencies operate in, and through, cities from London to Tokyo and Taipei, Budapest and Beijing to Sao Paulo. In particular, we have highlighted the role of territorial and network assets, their interdependencies and the geographically contingent 'coming together' of such assets that define the role of a city in advertising globalization. In this concluding chapter we, therefore, explore in more detail the impact of our analysis for both the academy and policy audiences.

Conceptual contributions

The analysis presented in this book has drawn inspiration from, and contributed to, work on knowledge-intensive business service firms and their globalization, and understandings of the world city. We have paid particular attention to the way our theoretical framing of the role of territorial and network assets can be used to explain asymmetries in the role of different cities in advertising globalization, the way global advertising agencies are embedded in cities and the different experiences of cities of the recession of 2009. As such, there are three key lines of conceptual contribution that the book makes to contemporary understandings of globalization and world cities.

Embedded and relational knowledge- intensive (advertising) business services

Throughout the book we have focused on the work of global advertising agencies and how, as a result of the changing role of advertising, the organization of work in agencies has evolved over time. In particular, we emphasized the implications of post-Fordist consumer cultures for the organization of worldwide campaigns, noting the switch from 'global' command and control campaigns, where adverts are developed centrally and rolled out in multiple markets, to transnational 'collaborate and cooperate' campaigns, where multiple campaigns are developed to respond to spatially entangled consumers. This switch and the drivers are exemplary of what in existing literature has been referred to as the embeddedness of global knowledge-intensive business service firms. Specifically, the discussion of global advertising agencies reveals that three forces lead to agencies being territorially embedded (Hess 2004) in certain cities:

1 The client. Whilst arguments about the perishability of services and the need for their production and consumption in close proximity are clearly open to debate (see for example, Bryson *et al.* 2004), and whilst our analysis of Detroit, and the out migration of advertising work there, suggests that agencies don't always need to locate in proximity to their clients, overall the importance of locating in a place where demand is high has been reaffirmed. In particular, a large and diverse pool of clients has been shown to embed firms in particular places and act as the basis for a resilient or adaptable advertising economy.
2 The consumer. As a cultural industry, advertising agencies are embedded by the importance of their relationship with the consumers targeted in adverts. Sharing the 'lifeworld' (Aspers 2010) of spatially entangled (Pike 2009) consumers is vital and, as a result, agencies continue to need to locate in proximity to large and diverse consumer audiences. The development of virtual forms of user-led innovation appear to have had a limited impact on this process, although may grow in significance in the future as Web 2.0 and 3.0 generate more dynamic forms of virtual interaction.
3 The project ecology. Because of the diversity of knowledge-bases needed to develop effective advertising campaigns, advertising agencies are embedded in cities that have deep pools of talented workers and sophisticated project ecologies. This allows both knowledge spillovers through interactions between workers in the same and different industries but also the assembling of temporary project teams that can be used to complete the work associated with a particular campaign (see for example, Grabher 2001; Sydow and Staber 2002).

Indeed, global advertising agencies have become more geographically embedded over time as a result of the increased reflexivity, individuality and spatial entanglement displayed by consumers. Whilst the rationale for opening overseas offices has long been explained with reference to location-specific advantages, such as the access to markets facilitated (Dunning and Norman 1987), the changes associated with a switch to post-Fordist models of advertising has made geographical embeddedness and the exploitation of place-specific territorial assets more important than ever. The main lesson for work on knowledge intensive business services is that processes of globalization have not led to the 'end of geography' (O'Brien 1992) or the 'death of distance' (Cairncross 1997).

Rather, globalization has, instead, led to a continuation of the importance of place. This idea is reinforced by considering the main research findings of the book in terms of the relational form of global advertising agencies.

One of the main themes of analysis in the book is the changing power and politics associated with the management of worldwide campaigns in global agencies. The demise of 'global' strategies and the rise of 'transnational' modes of organizing campaigns have led to subtle changes in the spatial relationships between agencies' offices. Collaboration and cooperation increasingly define relations because of the need to develop campaigns targeted at situated consumer groups, something that is only possible when the expertise and insight of personnel co-located with consumers is tapped into through worldwide office networks. Moreover the client base of global agencies is becoming increasingly geographically dispersed with both the home country of clients and their global or regional marketing managers increasingly located away from the twentieth century heartlands of advertising – the USA and Western Europe. This has meant more offices in agencies' networks have become strategic sites for advertising work, rather than just being post box offices responsible for adapting campaigns developed elsewhere. This has generated new work – such strategies are not zero sum games with one office losing work to another, but instead involve generating additional work as, for example, multiple offices produce multiple campaigns for one product rather than relying on a single worldwide advert – and has resulted in offices in South America (for example, Sao Paulo) and Asia (for example, Bangkok) growing in power.

Hence global advertising agencies require an effectively embedded relational organizational form with the success of the agency depending on collaboration and cooperation between multiple situated offices. Reflecting ideas developed in relational and global production network (GPN) approaches (Boggs and Rantisi 2003; Coe *et al.* 2008; Dicken *et al.* 2001; Yeung 2005), the analysis in this book suggests that it is vital to study knowledge-intensive business service firms through a lens that reveals both their situatedness – the way they are embedded in cities and the localization and agglomeration economies found there – but also their relatedness – the way flows of work, knowledge and people generate and allow the execution of global work. Global advertising agencies, and knowledge-intensive business service firms more generally, are, then, intensely territorial and network embedded organizations (see for example, Hess 2004).

Such multidimensional embeddedness can be best seen in relation to the way learning and innovation occurs in global advertising agencies. Being territorially embedded brings benefits in terms of the understandings of consumers gained and the access provided to talented workers that form a city's project ecology. Both allow the development of innovative campaigns. At the same time network embeddedness ensures learning and innovation occurs in stretched, often virtual spaces – for example when advertisers in different offices cooperate and collaborate using a combination of email, telephone and videoconference – as well as through to the mobility of talented workers across agencies' office networks.

It is, however, interesting to note that the nature of, in particular, the network embeddedness of global advertising agencies has been affected by the credit crisis and recession. The resurgence in popularity of the 'global' model in 2009 because of the cost savings it offers when running a worldwide campaign highlights the constantly shifting nature of spatial relationship in global firms. At one level this shows how the clients of agencies would indeed prefer a world in which geography and distance are irrelevant. Although this is not the case and advertisers would prefer to leave the global campaign in the past, budgetary constraints have certainly been driving geographical compromises. At another level, and perhaps most significantly, the resurgence of the global campaign highlights how the spatial power relations that exist in firms' worldwide office networks have been reconfigured again by the recession, in some cases leading to the status quo of old being reinstated as, for example, New York regains its 'command and control' role. However, as also noted, in some cases new players have emerged and have developed resources that allow them to compete with incumbents such as New York. Offices in cities such as Singapore are just as likely to take on 'global' leadership roles during the recession, something that would not have been the case in the latter years of the twentieth century.

Through our discussions in this book, we have, then, highlighted the spatial complexities of the contemporary global agency. We believe these complexities are relevant to the analysis of the work of all global knowledge-intensive business service firms, with the caveat that each particular type of business service will have subtly different forms of territorial and network embeddedness. We emphasize the importance of recognizing such spatial complexities because of their central role in also explaining the *city* geographies of (advertising) globalization.

(Advertising) cities in geographies of globalization

The territorial embeddedness of global advertising agencies reveals that cities continue to play a central role in advertising globalization. In particular, it is the mediation of processes of advertising work through certain cities that is one of the key themes of the book. A number of key insights have been developed in this regard.

1 Cities and territorial assets. The research has revealed that the place-specific assets of a city, whether they relate to clients, consumers or project ecologies, are key in determining its role in advertising globalization. At one level this relates to agglomeration and localization advantages. Successful cities are those cities that house a diverse array of industries with related but also unrelated variety (see for example, Boschma 2005; Glaeser *et al.* 1992; Jacobs 1969, 1984) because of the way the co-presence of an array of industries generates knowledge spillovers and project ecologies. At another level successful cities are also those with large consumer audiences that can be studied and targeted in campaigns, something intimately tied to the presence of a large and diverse array of industries that provide employment. These employers also act as the clients of agencies. But it is not just these territo-

rial assets that are key to a city's strategic importance in advertising globalization.

2 Cities and synergistic territorial and network assets. Most importantly it has been shown that it is the combination of a range of territorial *and* network assets that helps ground advertising globalization in a city. It is not enough to simply have a client base and demand; Detroit proves this point. Similarly it is not sufficient to just be a city with a large consumer market; Philadelphia proves this point (see Table 4.5). Nor is it enough to be a creative city with talented workers; Montreal proves this by its absence from Table 4.5. Instead it is the way all three of these assets combine to generate an urban milieu that is ideal for advertising work, and the way such milieus are simultaneously also generated by and generate network assets in the form of inwards and outward flows of trade and knowledge, which is key for the success of a city. For example, New York City is such a strategic site in advertising globalization because of its territorial assets but also the way these generate network assets (leadership of and collaboration on worldwide campaigns). But these territorial assets are also themselves generated by network assets (inward flows of knowledge and workers; advertising work resulting from collaboration which sustains the city's project ecology) (see Figure 6.2). As a result it is the 'external' (network) spaces of flows a city is part of and not just the 'internal' (territorial) advantages of a city that defines economic success.

3 In times of economic instability, such as the credit crisis and recession of 2007 onwards, cities with the most effective synergies between territorial assets and network assets are likely to be resilient, with others having to adapt (if assets are strong enough to allow this) or risk becoming vulnerable to decline.

The research findings reported help, therefore, to explain why cities' network relations are constantly in flux. Since the seminal work of Castells (2000) the importance of inter-city networks has been widely accepted, something work on world cities has further reinforced (see for example, Beaverstock *et al.* 2000; Sassen 2006a; Taylor 2004, 2009; Taylor *et al.* 2010, in press). Here we have added sophistication to such debates by showing the way such networks evolve over time in terms of their functionality and the politics and power of the flows within them. As the strategies of global agencies have changed so has the role of different cities in advertising globalization because of the way the ability of each city's assets to render the city more or less powerful also changes. For some this has been about a decline in powerfulness (for example, Detroit) or adaptation (for example, Los Angeles). For others it has been a story of evolution but resilience (New York City) or ascendency and empowerment (for example, Bangkok). But, in all cases, power relations change over time and new cities become powerful as new resources are generated as a result of changing economic conditions and associated effects on territorial and network assets. The growth in importance of offices in cities such as Bangkok and Sao Paulo is exemplary of such additionality which, whilst changing the role of offices in New York City, has not necessarily 'stolen' power from them. Rather it has led to New York City being powerful for different reasons – i.e. its command and control role but also its new cooperative role on campaigns managed by other offices.

Such additionality emphasizes, then, the fact that cities cooperate as much as they compete and that world cities are not just 'command and control' posts (Sassen 2000; Taylor 2004). The discussion in this book shows that even an apparently hegemonic world city such as New York, often assumed to be, along with London and Tokyo, *the* centre of power in the global economy, relies for its success as much on cooperation with other world cities as it does on its command and control role. Geographical contingencies determine the extent to which a city benefits from both command and control and cooperative roles in processes of (advertising) globalization with such contingencies being the subject of the research findings reported in this book.

In the context of our three case study US cities, this has meant that, to put it crudely, work that used to be exported as part of the command and control role in advertising globalization of US cities, is in many cases now produced 'locally' for overseas markets: for Jacobs (1969, 1984) this is a classic import replacement process creating 'new work' and therefore economic expansion. Two key trends are relevant in explanations of this change in spatial relationships and the related change in the territorial and network assets of different cities. First, the skill-base in advertising in 'developing' countries is improving rapidly and being called on to develop advertising that responds to spatially entangled post-Fordist consumer cultures. As a result, no longer is it the case that the US offices of global agencies 'teach' the world to do advertising. Rather, US offices are now part of a worldwide intra-firm collaboration that drives advertising innovation because of both changing network processes (the growth in the importance of the transnational campaign) and territorial processes (the development of skill sets in new cities). As reported previously, this mirrors the arguments of Speece *et al.* (2003) who point to the way Vietnam's advertising industry is changing from a situation where adverts are imported, to one where expertise gained from imported adverts and collaborations with global firms allows the development of locally competitive products. Similarly Po (2006) highlights how global agencies in Beijing, Guangzhou and Shanghai are not simply reproducing US adverts overseas, but, instead, are developing local models of advertising, something that newly emerging domestic agencies are quickly replicating so as to compete with global agencies (see also Muller 2005).

Second, reinforcing this trend and one of the most important future trends is the development of city client bases in Asia. Of particular significance is the emergence of market leaders in high value consumer goods from outside the USA and Western Europe. This inevitably drives the growth of offices, client management and strategic work in cities outside 'incumbent' locations such as New York, but also changes the nature of work in the 'incumbent' cities as again territorial processes (the development of important client bases in new cities) and network processes (the fact that global agencies need to develop cooperation between offices to service new clients based, for example, in Asia) reconfigure the roles of cities in advertising globalization. Hence it seems unthinkable, even in the context of the recession, that there will be a return to the days of US offices acting as the sole leaders of global advertising accounts. The 'transnational' campaign model has, therefore, led to a decline in global work in some US cities. But the process also benefits some US cities: importing advertising to the USA from elsewhere is not a strategy used because of the need to tailor adverts to the specificities of the internally differentiated US consumer market. There are, of course, other 'threats' to the US cities we studied. For example, their role relies on the strength of the US economy and consumer purchasing power. At the same time, with China competing for the title of the largest consumer market in the world, the USA will not be the only preoccupation of clients in the future. Nevertheless, it is hard to envisage a future where effectively tapping into the US market, through US based strategic and creative work, is not important and will not influence the geographies of advertising globalization. But, this will only occur in, and through, cities with the types of territorial and network assets discussed in the preceding arguments, i.e. those cities able to generate, and be generated by, network assets.

Changes in the geographies of (advertising) globalization have, then, resulted not from one city being outcompeted by another city but from a broader array of complex economic processes that have both generated new territorial assets for cities such as Bangkok and reconfigured the nature of the spaces of flows that make up 'incumbent' cities' network assets. The discussions throughout the book provide a way of further teasing out the effects of such complex factors and the way they define the role of different cities as sites of work in the global space economy. In doing this, the book helps to further grapple with contemporary conceptual conundrums associated with social science research of globalization.

Bringing work on business services and world cities into dialogue

An important debate has begun recently about how different conceptualizations, and specifically GPN (see for example, Dicken *et al.* 2001; Coe *et al.* 2008) and world city (see for example, Beaverstock *et al.* 2000; Sassen 2006a; Taylor 2004) discourses, can be brought into dialogue to more effectively theorize the way firms and cities operate in the contemporary period of economic globalization. Discussions about the pros and cons of such a meeting have been diverse and partly fractious (see for example, Brown *et al.* 2010; Coe *et al.* 2010; Sassen 2010; Weller 2008) and here we focus only upon how this book might contribute to one element of these debates.

Despite the risk of simplifying what are more complex theoretical issues, here we characterize one of the critiques that a dialogue between GPN and world cities discourses is supposed to resolve as follows: world cities literature lacks detail on network process (i.e. what flows between cities and what constitutes power relations) whilst global production network literatures fail to fully explore the places in which networks are embedded (i.e. the role of the cities that networks operate through) (see for example, Carroll 2007; Weller 2008). Proponents of both approaches can defend such accusations or explain the reasons for such superficiality but here, and building on this research, we suggest that a dialogue between the GPN and world cities approaches may well be fruitful in overcoming such limitations. Specifically in this book, we have shown how studying the work *processes* in global advertising agencies and the influences on these processes can help simultaneously explain:

1 The way agencies operate as relational network organizations and the role of the 'materiality' of the city and its assets in the embedding of these operations – i.e. we provide insights relevant to a GPN approach that overcome existing limitations.
2 What flows in world city networks, the factors that determine the nature of the network flows a city is connected into and the feedbacks in terms of the politics and power generated – i.e. we provide insights relevant to work on world cities that overcomes the limitations of the approach highlighted above.

To make such advances the book does two things. First, it takes inspiration from both the GPN and world cities literatures, and combines the insights of the two to make sense of the empirical conundrum that is the geographies of advertising globalization. As such the book is a theoretically informed empirical analysis rather than a theoretical project per se. Second, the book uses the insights from these two approaches to develop what Sassen (2006b) would call a 'focus on the non x'. Specifically, we do not focus our analysis solely on the cities or firms we are attempting to understand. Rather we focus on the 'non x', which for us is the advertising work process. By studying the work process we gain insights into the agencies and cities we want to understand, and, as the two previous arguments reveal, we produce insights that can help develop both the GPN and world cities approaches, because our analysis draws on, but is not bound exclusively by, the theoretical blinkers of one or another of the approaches. Specifically our analysis shows that:

1 World city network flows involve both trade and knowledge, are unidirectional, evolve over time, are powerful in ways determined by the influence of the interdependent territorial assets of the cities, and are essential if a city is to be a site of resilient, adaptable and regenerative economic activity.
2 GPNs are grounded in and produce cities in geographically contingent ways, are reliant on the territorial assets of particular cities, have relationships with cities that evolve over time and are political as a result of the way different cities' territorial assets allow them to develop more or less powerful spatial relationships.

Of course, some would argue that the theoretical foundations of the GPN and world cities approaches are incompatible and an attempt to contribute to both should not be made through a single piece of research. There may well be some validity in this argument. But as an attempt to deal with an empirical conundrum through a theoretically informed analysis we believe the approach taken here is effective and defensible.

Future research directions

As a result of the analysis presented in this book, a number of questions for future research emerge relating to advertising agencies, cities and the interactions between the two. Here we identify what we think are three particularly important questions. First, and in relation to agencies, the future of the global agency needs more focused analysis. Leslie (1997b) noted that the role and value of global agencies came under scrutiny as a result of shifts to post-Fordist consumer cultures and, in many ways, through transnational campaign models global agencies managed to respond to this crisis. Yet questions continue to exist about the synergies, cost savings or strategic benefits in terms of branding gained from using a global agency rather than multiple national agencies. The partial resurrection of global campaign models as a result of the current recession suggest that there will always be a place for economies of scale and global agencies are ideally placed to deliver such benefits to clients. But more and more the producers of consumer goods openly question whether the benefits of global agencies are outweighed by the costs. Our interviewees were all too aware of the common criticisms: global agencies are too bureaucratic, something that stifles creativity; billing models are opaque and, in the eyes of the client, used to hide the cost of the overheads associated with running a global agency; transnational collaboration is not as effective as it should be. Indeed, in relation to the final point, the Chief Executive of WPP, Sir Martin Sorrell, has regularly admitted in radio interviews and in the press that improving collaboration and cooperation between agencies' offices is vital for the future of the global agency model.

It would, therefore, seem important to look in more detail at how global agencies are reforming themselves in the twenty-first century and at the implications for the spatial relationships between offices. In this book we have documented the rise of the transnational model and its implications. A starting point for future research would be to focus on the functioning of this model from the perspective of non-US, non-Western European offices. For example, understanding the role, effectiveness and problems with the model from the perspective of those working in offices in Sao Paulo or Bangkok may reveal more insights into the way power relations enable or hinder collaboration and cooperation. Further lines of research might also consider the role of alternative models, for example segmentation through forms of relational proximity and grouping together offices serving consumer markets with shared characteristics, as highlighted by the analogy of Mattelart (1991) discussed earlier in the book that draws attention to the similarities between consumer groups located in parts of Paris and New

York City. Or research might examine the socio-technical architecture of effective collaborative and cooperative global agencies. For example, different uses of technology, business travel, expatriation and intra-organizational temporary teams whereby talent is managed agency wide rather than at the office level, and project teams assembled using personnel from throughout the world, could all be studied.

Second, and related to the first line of future research discussed above, the resilience of the boutique agency and the way they continue to compete with global agencies deserves further attention. One of the unexpected findings of our research was that global agencies feared boutique competitors more than other global agencies and increasingly suffered from key accounts being lost to boutiques – e.g. Coca-Cola® has moved a number of its accounts to independent agencies. And even more surprisingly these boutiques are not always located in advertising hotspots such as New York City. Interviewees gave examples of accounts moving from our three case study cities to places including Denver, Portland, Minneapolis and Dallas. Understanding how the organizational structures of such boutique agencies allows them to compete with global agencies would seem like an important priority, as would understanding how such agencies operate in cities not renowned for their advertising related territorial or network assets. Just how do boutiques located in, for example, Portland manage to attract clients, talented workers and form temporary project teams? What effects does the success of such boutiques have on a city's territorial and network assets? To what extent can boutiques, that often only have one office, engage in global work and manage campaigns running in multiple regions or countries?

Third, it would also seem somewhat urgent to translate the analysis presented here of how global agencies operate in and through US world cities into analysis of emerging cities in South America, Asia and the Middle East. As we argued at the start of the chapter, the processes we have identified in this book are relevant to advertising globalization throughout the world and not just in the USA. However, we also recognize the important geographical contingencies that determine the territorial and network assets of a city and their affects on a city's role in (advertising) globalization. It would, thus, seem valuable to explore the relationships between global agencies and the client bases, project ecologies and consumer audiences in a range of cities in emerging markets and simultaneously explore the type of network assets and spatial relationships generated. This would also relate to our first agenda for future research and fully take account of the decentring that is occurring in processes of globalization as a multi-polar world emerges in which the USA and Western Europe coexist with other important centres of economic activity.

Policy implications

The implications of our analysis for policy primarily relate to cities and the 'management' of their role in the global space economy. Specifically, we are able to make two key and interrelated points about the factors determining the success of a city in the contemporary space economy.

1 Cities are complex and synergistic. The discussion of the way territorial and network assets combine to determine the resilience, adaptability and general strength of a city's (advertising) economy reveals that focusing on a one dimensional strategy as part of attempts to sustain or regenerate a city is likely to lead to failure. Successful cities are those than combine territorial assets that create demand (a market) with assets that ensure supply (workers capable of serving the market) with spillovers between related and unrelated sectors that reproduce both demand and supply.
2 Successful cities are also networked cities. Cities that are able to manufacture themselves a strategic role in inter-city, often global, networks of trade and knowledge are less vulnerable and more likely to prosper as economic demand and work processes change. But the networks that allow such resilience and prosperity are produced by and produce territorial assets. As such, the challenge is to find a way to generate a more holistic city that has both territorial and network assets in order to generate an economically sustainable city. Such cities have been shown to be more resilient and adaptable (e.g. New York), helping avoid vulnerability during times of economic change.

Perhaps the best way to assess the implications of these findings for policy practices is to contextualize them in relation to existing debates about urban regeneration. We choose one key approach to frame our discussion here.

Through his original book (Florida 2002) and subsequent articles (for example, Stolarick and Florida 2006), Richard Florida has developed a treatise about the role of the creative class in driving urban regeneration. Under the guise of the 'Three Ts' model of the role of 'technology, talent and tolerance', the creative class thesis posits that cities seeking to rejuvenate their economies should attract technologically innovative industries and talented workers and generate an environment tolerant of diversity. As a result, many cities have pinned their hopes on new science parks and public art as mechanisms to develop a creative economy. Florida's ideas have been critiqued from many angles, with accusations existing about the tautology of his thesis and the fact that it has no answer to the question of whether economic activity precedes the arrival of the creative class or results from their arrival (see for example, Peck 2005; Pratt 2008). There have also been critiques relating to the social consequences for the 70 per cent of the population who are not creative workers if all cities strive to be creative cities and prioritize the demands of the elite creative class (see, for example, Markusen 2006; Peck 2005). Our analysis of the role of cities in advertising globalization and the two key policy relevant findings summarized above lend weight to the former critique in particular whilst not undermining the latter.

The analysis in this book reveals a major flaw in urban regeneration plans that prioritize attracting new creative workers to a city. Such workers only guarantee economic prosperity if the wider structural demand and supply conditions exist that will support economic activity. In the case of advertising, attracting talented advertising workers to the city will not create a role for the city in advertising markets nationally or globally if: (1) there are not clients in the city demanding the services of those advertisers; (2) there is not a large and diverse consumer market in the city that advertisers can study and share a space with; (3) the related and unrelated industries that make up the advertising project ecology are not co-present in the city; and (4) points (1) to (3) don't lead to the city having important network connections, nationally and internationally. It is the synergies and multiplier effects that cities generate that leads to economic success and often these synergies develop over many years – coincidence is, for example, widely accepted as the explanation for Silicon Valley's economic success. Consequently, a policy that focuses on attracting a particular type of creative worker is unlikely to generate the synergies needed for economic growth.

Florida has himself begun to recognize the role of such synergistic effects and also the way history – i.e. path dependence – can determine the success of a city. In Florida *et al.* (2010: 786) it is, for example, acknowledged that cities with successful music industries are those that combine 'economies of scale and economies of scope'. Scale relates to demand – successful cities are those with consumer markets that generate constant demand for music events – whilst scope relates to the project ecology of music – successful cities are those with related industries that allow the entire music production process to occur in one city. The explanation for cities with both scale and scope in the period studied (1970 to 2000) is, according to Florida *et al.* (2010: 789), 'a function of their past location and the past location of related cultural industries'.

So does this mean that processes of economic development in cities are unmanageable? To a certain extent, yes. Organic growth is most likely to lead to synergy and coincidence, which are hard to manufacture and control. But, another way of looking at the challenge is to consider the research findings reported in this book and the way they identify the need for a strategy that has multidimensionality. Debates about entrepreneurial cities and the role of government in driving growth (see for example, Hall and Hubbard 1998; Harvey 1989b; Jessop 2000) are useful to frame such a discussion. Recently developed by Golubchikov (2010) to conceptualize the idea of world city entrepreneurialism, studies of entrepreneurial cities have revealed how the role of government has slowly changed to be understood, as Harvey (1989b) notes, to involve: (1) attracting flows of capital by enhancing city assets; (2) acquiring command and control functions in the work of firms; (3) generating consumer demand by providing an attractive living environment; (4) redistributing the benefits of growth. Again there are many critiques of such approaches, particularly in relation to the decline in the role of government in managing social welfare and the regular failure of policies designed to incentivize people and firms to move to a city. But such approaches do appear more in line with the interpretation of the factors determining the success of a city developed in this book. Combining all four components of the entrepreneurial cities approach is more likely to generate demand and supply territorial assets, as well as network assets, and potentially the synergies and coincidences that spin off from them. As Chapter 9 suggests, when there is economic change, such as a recession, a city built on a multidimensional policy is more likely to be resilient or able to adapt to new demand and supply issues.

But, this does not dispel the challenge of actually designing, funding and implementing such a multidimensional approach whilst also ensuring social equality. It does not necessarily resolve the tautology of what/who comes first, demand or supply. As such the analysis in this book identifies and frames the complex multidimensional issues that need to be considered in any urban (re) development strategy. But, in doing so, the book also highlights the importance of existing critiques of city management policies and sets the scene for future research that focuses on the challenges of synergy, coincidence and the what/who comes first question.

Appendix

Profile of interviewees

ID	*Position*	*City*
1	Vice-President	New York
2	Director of Insights and Brand Planning	New York
3	Global Chief Strategy Officer	New York
4	Behavioral Planning Director	New York
5	*CEO*	*New York*
6	*Director of Engagement Strategy*	*New York*
7	Chief Creative Officer	New York
8	Head of Multinational Accounts	New York
9	Director of Trendspotting	New York
10	Head of Planning	New York
11	Vice-President	New York
12	CEO	New York
13	CEO	New York
14	Editor	New York
15	*Research and Planning Director*	*Detroit*
16	Executive Creative Director	Detroit
17	Director USA Broadcast Promotion	Detroit
18	President	Detroit
19	President	Detroit
20	Vice-President	Detroit
21	International Strategic Planning, Analyst	Detroit
22	Executive Creative Director	Los Angeles
23	Director of Account Planning	Los Angeles
24	General Manager	Los Angeles
25	Chief Creative Officer	Los Angeles
26	Director of Account Planning	Los Angeles
27	Chief Strategy Officer	Los Angeles
28	Chairman, CEO	Los Angeles
29	Partner/Account Director	Los Angeles
30	*Director of Integrated Insights*	*Los Angeles*

Note
Lines in italics indicate an interviewee that participated in follow-up interviews about the effects of the recession on different advertising cities.

Bibliography

Aaker, D.A. and Jacobson, R. (2001) 'The value relevance of brand attitude in high-technology markets', *Journal of Marketing Research*, 38: 485–93.

Advertising Age. (2000) 'Special report. The global landgrab: its now or never', *Advertising Age*, 16 June.

—— (2002) 'Annual agency report 2002', New York, *Advertising Age*, 22 April.

—— (2009) 'Agency Report 2009', New York, *Advertising Age*, 27 April.

—— (2010a) 'Agency Report 2010', New York, *Advertising Age*, 25 April.

—— (2010b) 'Auto marketing accounts detour from Detroit: could Portland, Minneapolis, Seattle or Dallas emerge as the New Motor City', *Advertising Age*, 4 January.

—— (2010c) *US ad industry employment 2000–2010*, *Advertising Age* Data Center. Available at http://adage.com/datacenter (accessed 1 September 2010).

—— (2010d) *US measured ad spend by category*, *Advertising Age* Data Center. Available at http://adage.com/datacenter (accessed 1 September 2010).

Allard, S., Tolman, R. and Rosen, D. (2003) 'The geography of need: spatial distribution of barriers to employment in metropolitan Detroit', *Policy Studies Journal*, 31: 293–308.

Allen, J. (1992) 'Services and the UK space economy: regionalization and economic dislocation', *Transactions of the Institute of British Geographers*, 17: 292–305.

—— (1995) 'Crossing borders: footloose multinationals', in Allen, J. and Hamnett, C. (eds) *A shrinking world? Global unevenness and inequality*. Oxford: Oxford University Press.

—— (2000) 'Power/economic knowledge: symbolic and spatial formations', in Bryson, J., Daniels, P.W., Henry, N. and Pollard, J. (eds) *Knowledge, space, economy*. London: Routledge.

—— (2002) 'Living on thin abstractions: more power/economic knowledge', *Environment and Planning A*, 34: 451–66.

—— (2003) *Lost geographies of power*. Oxford: Blackwell.

Alter, S. (1994) *Truth well told: McCann-Erickson and the pioneering of global advertising*. London: McCann-Erickson Worldwide Publishers.

Alvesson, M. (2001) 'Knowledge work: ambiguity, image and identity', *Human relations*, 54: 863–86.

—— (2004) *Knowledge work and knowledge intensive firms*. Oxford: Oxford University Press.

Amin, A. (2002) 'Spatialities of globalization', *Environment and Planning A*, 34: 385–99.

Amin, A. and Cohendet, P. (1999) 'Learning and adaptation in decentralised business networks', *Environment and Planning D: Society and Space*, 17: 87–104.

Amin, A. and Cohendet, P. (2004) *Architectures of knowledge: firms capabilities and communities*. Oxford: Oxford University Press.
Amin, A. and Thrift, N. (2002) *Cities: reimagining the urban*, Cambridge: Polity Press.
Amin, A. and Thrift, N.J. (1994) 'Globalisation, institutional "thickness" and the local economy', in Healey, P., Cameron, S., Davoudi, S., Graham, S. and Madanipour, A. (eds) *Urban management*. London: Belhaven.
Arvidsson, A. (2006) *Brands: meaning and value in media culture*. London and New York: Routledge.
Asheim, B., Coenen, L. and Vang, J. (2007) 'Face-to-face, buzz, and knowledge bases: sociospatial implications for learning, innovation, and innovation policy', *Environment and Planning C: Government and Policy*, 25: 655–70.
Aspers, P. (2010) 'Using design for upgrading in the fashion industry', *Journal of Economic Geography*, 10: 189–207.
Bagchi-Sen, S. and Sen, J. (1997) 'The current state of knowledge in international business in producer services', *Environment and Planning A*, 29: 1153–74.
Balmer, J. and Greyser, S. (2003) *Revealing the corporation: perspectives on identity, image, reputation, corporate branding, and corporate-level marketing: an anthology*. London: Routledge.
Bartlett, C. and Ghoshal, S. (1998) *Managing across borders: the transnational solution*. London: Random House.
Bathelt, H. and Glückler, J. (2003) 'Toward a relational economic geography', *Journal of Economic Geography*, 3: 117–44.
Bathelt, H., Malmberg, A. and Maskell, P. (2004) 'Clusters and knowledge: local buzz, global pipelines and the process of knowledge creation', *Progress in Human Geography*, 28: 31–56.
Beaverstock, J.V. (1996) 'Subcontracting the accountant! Professional labour markets, migration, and organisational networks in the global accountancy industry', *Environment and Planning A*, 28: 303–26.
—— (2004) '"Managing across borders": knowledge management and expatriation in professional legal service firms', *Journal of Economic Geography*, 4: 157–79.
—— (2007a) 'Transnational work: global professional labour markets in professional service accounting firms', in Bryson, J. and Daniels, P.W. (eds) *The Handbook of Service Industries*. Cheltenham: Edward Elgar.
—— (2007b) 'World city networks from below: international mobility and inter-city relations in the global investment banking industry', in Taylor, P.J., Derudder, B., Saey, P. and Witlox, F. (eds) *Cities in globalization: practices, policies, theories*. London: Routledge.
Beaverstock, J.V., Smith, R. and Taylor, P.J. (1999a) 'The long arm of the law: London's law firms in a globalising world economy', *Environment and Planning A*, 13: 1857–76.
Beaverstock, J.V., Smith, R. and Taylor, P.J. (1999b) 'A roster of world cities', *Cities*, 16: 445–58.
Beaverstock, J.V., Smith, R. and Taylor, P.J. (2000) 'World-city network: a new metageography', *Annals of the Association of American Geographers*, 90: 123–34.
Beaverstock, J.V., Derudder, B., Faulconbridge, J.R. and Witlox, F. (2009) 'International business travel: some explorations', *Geografiska Annaler: Series B, Human Geography*, 91: 193–202.
Benner, C. (2003) 'Learning communities in a learning region: the soft infrastructure of cross-firm learning networks in Silicon Valley', *Environment and Planning A*, 35: 1809–30.

Benneworth, P. and Henry, N. (2004) 'Where is the value added in the cluster approach? Hermeneutic theorising, economic geography and clusters as a multiperspectival approach', *Urban Studies*, 41: 1011–23.

Blackler, F., Crump, N. and Mcdonald, S. (1998) 'Knowledge, organizations and competition', in Krogh, G., Roos, J. and Kleine, D. (eds) *Knowing in firms: understanding, managing and measuring knowledge*. London: Sage

Boden, D. and Molotch, H. (1994) 'The compulsion of proximity', in Friedland, R. and Boden, D. (eds) *NowHere: space, time and modernity*. Berkeley, CA: University of California Press.

Boggs, J.S. and Rantisi, N. (2003) 'The relational turn in economic geography', *Journal of Economic Geography*, 3: 109–16.

Boschma, R.A. (2005) 'Proximity and innovation: a critical assessment', *Regional Studies*, 39: 61–74.

Brenner, N. and Keil, R. (eds) (2005) *The global cities reader*. London and New York: Routledge.

Brock, D.M., Powell, M.J. and Hinings, C.R. (1999) 'Restructuring the professional organization: corporates, cobwebs and cowboys', in Brock, D.M., Powell, M.J. and Hinings, C.R. (eds) *Restructuring the professional organization: accounting, healthcare and law*. London and New York: Routledge.

Brown, E., Derudder, B., Parnreiter, C., Pelupessy, W., Taylor, P.J. and Witlox, F. (2010) 'World city networks and global commodity chains: towards a world-systems' integration', *Global Networks*, 10: 12–34.

Brown, J.S. and Duguid, P. (1991) 'Organizational learning and communities-of-practice: toward a unified view of working, learning, and innovation', *Organization Science*, 2: 40–57.

Bryson, J.R. (2007) 'The second global shift: the offshoring or global sourcing of corporate services and the rise of distanciated emotional labour', *Geografiska Annaler: Series B, Human Geography*, 89: 31–43.

Bryson, J.R., Daniels, P.W. and Warf, B. (2004) *Service worlds*. London: Routledge.

Bunnell, T. and Coe, N. (2001) 'Spaces and scales of innovation', *Progress in Human Geography*, 25: 569–89.

Burrage, M., Jarausch, K. and Sigrist, H. (1990) 'An actor-based framework for the study of the professions', in Burrage, M. and Torstendahl, R. (eds) *Professions in theory and history*. London: Sage.

Cairncross, F. (1997) *The death of distance: how the communications revolution will change our lives*. London: Orion.

Carroll, W. (2007) 'Global cities in the global corporate network', *Environment and Planning A*, 39: 2297.

Castells, M. (2000) *The rise of the network society*. Malden, MA: Blackwell.

Castree, N. (2001) 'Commodity fetishism, geographical imaginations and imaginative geographies', *Environment and Planning A*, 33: 1519–26.

Clarke, D.B. and Bradford, M.G. (1989) 'The uses of space by advertising agencies within the United Kingdom', *Geografiska Annaler B: Human Geography*, 71B: 139–51.

CNN (2010) 'Assignment Detroit'. Available at http://money.cnn.com/news/specials/assignment_detroit (accessed 18 May 2010).

Coe, N.M., Dicken, P. and Hess, M. (2008) 'Global production networks: realizing the potential', *Journal of Economic Geography*, 8: 1–25.

Coe, N.M., Dicken, P., Hess, M. and Yeung, H.W. (2010) 'Making connections: global production networks and world city networks', *Global Networks*, 10: 138–49.

Cooke, P. and Lazzeretti, L. (eds) (2008a) *Creative cities, cultural clusters and local economic development*. Cheltenham: Edward Elgar.

—— (2008b) 'Creative cities: an introduction', In Cooke, P. and Lazzeretti, L. (eds) *Creative cities, cultural clusters and local economic development*. Cheltenham: Edward Elgar.

Cooper, D., Hinings, C.R., Greenwood, R. and Brown, J.L. (1996) 'Sedimentation and transformation in organizational change: the case of Canadian law firms', *Organization Studies*, 17: 623–47.

Corporation, C.O.L. (2009) 'The global financial centres index', London, City of London Corporation/ZYen Group.

Crang, P. (1996) 'Displacement, consumption, and identity', *Environment and Planning A*, 28: 47–68.

Crewe, L. and Lowe, M. (1995) 'Gap on the map? Towards a geography of consumption and identity', *Environment and Planning A*, 27: 1877–98.

Currah, A. and Wrigley, N. (2004) 'Networks of organizational learning and adaptation in retail TNCs', *Global Networks*, 4: 1–23.

Currid, E. (2006) 'New York as a global creative hub: a competitive analysis of four theories on world cities', *Economic Development Quarterly*, 20: 330–50.

Daniels, P.W. (1993) *Service industries in the world economy*. Oxford: Blackwell.

—— (1995) 'The internationalisation of advertising services in a changing regulatory environment', *Service Industries Journal*, 15: 276–94.

De Mooij, M.K. (2004) *Consumer behavior and culture: consequences for global marketing and advertising*. London: Sage.

—— (2005) *Global marketing and advertising: understanding cultural paradoxes*. London: Sage.

De Mooij, M.K. and Keegan, W. (1991) *Advertising worldwide: concepts, theories and practice of international, multinational and global advertising*. London: Prentice Hall.

Derudder, B., Taylor, P.J., Ni, P., De Vos, A., Hoyler, M., Hassens, H., Bassens, D., Huang, J., Witlox, F. and Yang, X. (2010) 'Pathways of change: shifting connectivities in the world city network, 2000–2008', *Urban Studies*, 47: 1861–77.

Dicken, P. (2003) *Global shift*. London: Sage.

—— (2007) *Global shift*. London: Sage.

Dicken, P., Kelly, P.F., Olds, K. and Yeung, H.W.-C. (2001) 'Chains and network, territories and scales: towards a relational framework for analysing the global economy', *Global Networks*, 1: 89–112.

Dunning, J. and Norman, G. (1987) 'Theory of multinational enterprise', *Environment and Planning A*, 15: 675–92.

Dwyer, C. and Crang, P. (2002) 'Fashioning ethnicities: the commercial spaces of multiculture', *Ethnicities*, 2: 410.

Dwyer, C. and Jackson, P. (2003) 'Commodifying difference: selling EASTern fashion', *Environment and Planning D: Society and Space*, 21: 269–91.

Eisinger, P. (2003) 'Reimagining Detroit', *City and Community*, 2: 85–99.

Ekinsmyth, C. (2002) 'Project organization, embeddedness and risk in magazine publishing', *Regional Studies: The Journal of the Regional Studies Association*, 36: 229–43.

Empson, L. (2001) 'Introduction: knowledge management in professional service firms', *Human Relations*, 54: 811.

Englis, B.G. (1994) *Global and multinational advertising*. Hillside, NJ: Lawrence Erlbaum Associates.

Engwall, M. (2003) 'No project is an island: linking projects to history and context', *Research Policy*, 32: 789–808.

Faulconbridge, J.R. (2004) 'London and Frankfurt in Europe's evolving financial centre network', *Area*, 36: 235–44.

—— (2006) 'Stretching tacit knowledge beyond a local fix? Global spaces of learning in advertising professional service firms', *Journal of Economic Geography*, 6: 517–40.

—— (2007a) 'Exploring the role of professional associations in collective learning in London and New York's advertising and law professional service firm clusters', *Environment and Planning A*, 39: 965–84.

—— (2007b) 'Relational spaces of knowledge production in transnational law firms', *Geoforum*, 38: 925–40.

—— (2008) 'Managing the transnational law firm: a relational analysis of professional systems, embedded actors and time-space sensitive governance', *Economic Geography*, 84: 185–210.

Faulconbridge, J.R. and Muzio, D. (2007) 'Reinserting the professional into the study of professional service firms: the case of law', *Global Networks*, 7: 249–70.

Faulconbridge, J.R. and Muzio, D. (2008) 'Organizational professionalism in globalizing law firms', *Work, Employment and Society*, 22: 7–25.

Faulconbridge, J.R., Hall, S. and Beaverstock, J.V. (2008) 'New insights into the internationalization of producer services: organizational strategies and spatial economies for global headhunting firms', *Environment and Planning A*, 40: 210–34.

Faulconbridge, J.R., Beaverstock, J.V., Derudder, B. and Witlox, F. (2009) 'Corporate ecologies of business travel in professional service firms: working towards a research agenda', *European Urban and Regional Studies*, 16: 295.

Fendley, A. (1995) *Commercial break: the inside story of Saatchi & Saatchi*. London: Hamish Hamilton.

Florida, R. (2002) *The rise of the creative class*. New York: Basic Books.

Florida, R., Mellander, C. and Stolarick, K. (2010) 'Music scenes to music clusters: the economic geography of music in the US, 1970–2000', *Environment and Planning A*, 42: 785–804.

Frenken, K., Van Oort, F. and Verburg, T. (2007) 'Related variety, unrelated variety and regional economic growth', *Regional Studies: The Journal of the Regional Studies Association*, 41: 685–97.

Friedman, J. (1986) 'The world city hypothesis', *Development and Change*, 17: 69–83.

Glaeser, E.L., Kallal, H.D., Scheinkman, J.A. and Shleifer, A. (1992) 'Growth in cities', *Journal of Political Economy*, 100: 1126–52.

Glaser, B. and Strauss, A. (1967) *The discovery of grounded theory*. Chicago, IL: Aldine.

Goffman, E. (1967) *Interaction ritual: essays on face to face behaviour*, New York: Doubleday and Company.

Golubchikov, O. (2010) 'World-city-entrepreneurialism: globalist imaginaries, neoliberal geographies, and the production of new St Petersburg', *Environment and Planning A*, 42: 626–43.

Grabher, G. (1993) *The embedded firm: on the socioeconomics of industrial networks*. London: Routledge.

—— (2001) 'Ecologies of creativity: the village, the group and the heterarchic organisation of the British advertising industry', *Environment and Planning A*, 33: 351–74.

—— (2002) 'Cool projects, boring institutions: temporary collaboration in social context', *Regional Studies*, 36: 205–14.

—— (2004) 'Learning in projects, remembering in networks? Communality, sociality

and connectivity in project ecologies', *European Urban and Regional Studies*, 11: 103–23.
Grabher, G. and Ibert, O. (2006) 'Bad company? The ambiguity of personal knowledge networks', *Journal of Economic Geography*, 6: 251–71.
Grabher, G., Ibert, O. and Flohr, S. (2009) 'The neglected king: the customer in the new knowledge ecology of innovation', *Economic Geography*, 84: 253–80.
Granovetter, M. (1985) 'Economic action and social structure: the problem of embeddedness', *American Journal of Sociology*, 91: 481–510.
Grein, A. and Ducoffe, R. (1998) 'Strategic responses to market globalisation among advertising agencies', *International Journal of Advertising*, 19: 301–19.
Guardian. (2010) 'Global trade slumped 12% last year', *Guardian*, 24 February.
Hackley, C. (1999) 'Tacit knowledge and the epistemology of expertise on strategic marketing management', *European Journal of Marketing*, 33: 720–35.
Hall, P. (1966) *The world cities*. London: Widenfeld and Nicolson.
Hall, T. and Hubbard, P. (1998) *The entrepreneurial city: geographies of politics, regime and representation*. Chichester: John Wiley and Sons.
Hardy, C. (1996) 'Understanding power: bringing about strategic change', *British Journal of Management*, 7: S3–S16.
Harvey, D. (1989a) *The condition of postmodernity: an enquiry into the origins of cultural change*. Malden, MA: Wiley-Blackwell.
—— (1989b) 'From managerialism to entrepreneurialism: the transformation in urban governance in late capitalism', *Geografiska Annaler. Series B. Human Geography*, 71: 3–17.
Held, D., Mcgrew, A., Goldblatt, D. and Perraton, J. (1999) *Global transformations: politics, economics and culture*. Cambridge: Polity Press.
Henry, N. and Pinch, S. (2000) 'Spatialising knowledge: placing the knowledge community of Motor Sport Valley', *Geoforum*, 31: 191–208.
—— (2001) 'Neo-Marshallian nodes, institutional thickness, and Britain's motor sport valley: thick or thin?', *Environment and Planning A*, 33: 1169–84.
Hess, M. (2004) 'Spatial relationships? Towards a reconceptualization of embeddedness', *Progress in Human Geography*, 28: 165–86.
Hobday, M. (2000) 'The project-based organisation: an ideal form for managing complex products and systems?' *Research Policy*, 29: 871–93.
Hong, J.F.L., Snell, R.S. and Easterby-Smith, M. (2006) 'Cross-cultural influences on organizational learning in MNCS: the case of Japanese companies in China', *Journal of International Management*, 12: 408–29.
Jacobs, J. (1969) *The economy of cities*. New York: Vintage.
—— (1984) *Cities and the wealth of nations*. New York: Vintage.
Jeppesen, L. and Molin, M. (2003) 'Consumers as co-developers: learning and innovation outside the firm', *Technology Analysis and Strategic Management*, 15: 363–83.
Jessop, B. (2000) 'Globalisation, entrepreneurial cities and the social economy', in Hamel, P., Lustiger-Thaler, H. and Mayer, M. (eds) *Urban movements in a globalising world*. London: Routledge.
Johns, J. (2006) 'Video games production networks: value capture, power relations and embeddedness', *Journal of Economic Geography*, 6: 151–80.
Johnson, S. (2002) *Emergence: the connected lives of ants, brains, cities and software*. London: Penguin.
Jones, A. (2002) 'The global city misconceived: the myth of "global management" in transnational service firms', *Geoforum*, 33: 335–50.

—— (2003) *Management consultancy and banking in an era of globalization*. Basingstoke: Palgrave Macmillan.

—— (2007) 'More than "managing across borders?" The complex role of face-to-face interaction in globalizing law firms', *Journal of Economic Geography*, 7: 223–46.

—— (2008) 'The rise of global work', *Transactions of the Institute of British Geographers*, 33: 12–26.

—— (2009) 'Theorizing global business spaces', *Geografiska Annaler: Series B, Human Geography*, 91: 203–18.

Keller, K.L. (1993) 'Conceptualizing, measuring, and managing customer-based brand equity', *Journal of Marketing*, 57: 1–22.

Kodras, J. (1997) 'The changing map of American poverty in an era of economic restructuring and political realignment', *Economic Geography*, 73: 67–93.

Kotabe, M. and Helsen, K. (2001) *Global marketing management* (2nd edn). New York: Wiley.

Kyser, J., Sidhu, N.D., Ritter, K. and Guerra, F. (2010) 'Entertainment and the media in Los Angeles', Los Angeles, LAEDC Center for Economic Research. Available at: http://laedc.org (accessed 1 September 2010).

Landry, C. (2000) *The creative city: a toolkit for urban innovations*. London: Earthscan.

Lash, S. and Urry, J. (1994) *Economies of signs and spaces*. London: Sage.

Lave, J. and Wenger, E. (1991) *Situated learning: legitimate peripheral participation*. Cambridge: Cambridge University Press.

Lawson, C. and Lorenz, E. (1999) 'Collective learning, tacit knowledge and regional innovative capacity', *Regional Studies*, 33: 305–17.

Lears, J. (1995) *Fables of abundance: a cultural history of advertising in America*. New York: Basic Books.

Leslie, D. (1995) 'Global scan: the globalization of advertising agencies, concepts and campaigns', *Economic Geography*, 71: 402–26.

—— (1997a) 'Abandoning Madison Avenue: the relocation of advertising services in New York city', *Urban Geography*, 18: 568–90.

—— (1997b) 'Flexibly specialized agencies? Reflexivity, identity, and the advertising industry', *Environment and Planning A*, 29: 1017–38.

Liu, W. and Dicken, P. (2006) 'Transnational corporations and "obligated embeddedness": foreign direct investment in China's automobile industry', *Environment and Planning A*, 38: 1229–47.

Los Angeles County Economic Development Corporation (2009) *OTIS Report on the Creative Economy of the Los Angeles Region*, Los Angeles County Economic Development Corporation, LAEDC, Los Angeles. Available at: www.LAEDC.org (accessed 18 May 2010).

Lury, C. (2004) *Brands: the logos of the global economy*. London: Routledge.

McCarthy, J. (1997) 'Revitalization of the core city: the case of Detroit', *Cities*, 14: 1–11.

McCracken, G. (1993) 'The value of the brand: an anthropological perspective', in Aaker, D.A. and Biel, A.L. (eds) *Brand equity and advertising*. Hillside, NJ: Lawrence Erlbaum Associates.

McNeill, D. (2008) *The global architect: firms, fame and urban form*. London and New York: Routledge.

Madden, T., Fehle, F. and Fournier, S. (2006) 'Brands matter: an empirical demonstration of the creation of shareholder value through branding', *Journal of the Academy of Marketing Science*, 34: 224–35.

Marcuse, H. (1964) *One dimensional man*. Boston, MA: Beacon Press.

Markusen, A. (2006) 'Urban development and the politics of a creative class: evidence from a study of artists', *Environment and Planning A*, 38: 1921–40.

Marshall, A. (1952) *Principles of economics (1890)*. London: Macmillan.

Marston, S., Jones, J.P. and Woodward, K. (2005) 'Human geography without scale', *Transactions of the Institute of British Geographers NS*, 30: 416–32.

Martin, R. and Sunley, P. (2003) 'Deconstructing clusters: chaotic concept or policy panacea', *Journal of Economic Geography*, 3: 5–35.

Maskell, P., Bathelt, H. and Malmberg, A. (2006) 'Building global knowledge pipelines: the role of temporary clusters', *European Planning Studies*, 14: 997–1013.

Mateos-Garcia, J. and Steinmueller, W.E. (2008) 'Open, but how much? Growth, conflict, and institutional evolution in open-source communities', in Amin, A. and Roberts, J. (eds) *Community, economic creativity, and organization.* Oxford: Oxford University Press.

Mather, O. (2010) 'Ogilvy & Mather'. Available at: www.ogilvy.com/About/Network/Ogilvy-Mather.aspx (accessed 24 May 2010).

Mattelart, A. (1991) *Advertising international: the privatisation of public space*. London and New York: Routledge.

Millar, J. and Salt, J. (2008) 'Portfolios of mobility: the movement of expertise in transnational corporations in two sectors – aerospace and extractive industries', *Global Networks*, 8: 25–50.

Miller, D. (1995) *Acknowledging consumption: a review of new studies*. London: Routledge.

Molotch, H. (1998) 'LA as design product: how art works in a regional economy', in Scott, A.J. and Soja, E. (eds) *The city: Los Angeles and urban theory at the end of the twentieth century*. London: University of California Press.

—— (2003a) 'Place in product', *International Journal of Urban and Regional Research*, 26: 665–88.

Molotch, H.L. (2003b) *Where stuff comes from: how toasters, toilets, cars, computers, and many other things come to be as they are*. London: Routledge.

Morgan, K. (2004) 'The exaggerated death of geography: learning, proximity and territorial innovation systems', *Journal of Economic Geography*, 4: 3–21.

Muller, L. (2005) 'Localizing international business service investment: the advertising industry in South East Asia', in Daniels, P.W., Ho, K.C. and Hutton, T.A. (eds) *Service industries and Asia-Pacific cities: new development trajectories*. London and New York: Routledge.

Nachum, L. (1999) *The origins of the international competitiveness of firms*. Cheltenham: Edward Elgar.

Nachum, L. and Keeble, D. (2000) 'Foreign and indigenous firms in the media cluster of central London', *ESRC Centre for Business Research Working Paper No. 154*.

Neill, W., Fitzsimons, D. and Murtagh, B. (1995) *Reimaging the pariah city: urban development in Belfast and Detroit*. Aldershot: Avebury.

O'Brien, R. (1992) *Global financial integration: the end of geography*. London: Royal Institute of International Affairs.

O'Farrell, P.N. and Hitchens, D.M. (1990) 'Producer services and regional development: key conceptual issues of taxonomy and quality measurement', *Regional studies*, 24: 163–71.

Orlikowski, W. (2002) 'Knowing in practice: enacting a collective capability in distributed organizing', *Organization Science*, 13: 249–73.

Papavassiliou, N. and Stathakopoulos, V. (1997) 'Standardization versus adaptation of international advertising strategies', *European Journal of Marketing*, 31: 504–27.

Peck, J. (2005) 'Struggling with the creative class', *International Journal of Urban and Regional Research*, 29: 740–70.
Perry, M. (1990) 'The internationalisation of advertising', *Geoforum*, 21: 33–50.
Pike, A. (2009) 'Geographies of brands and branding', *Progress in Human Geography*, 33: 619–45.
Po, L. (2006) 'Repackaging globalization: a case study of the advertising industry in China', *Geoforum*, 37: 752–64.
Polanyi, K. (1944) *The great transformation*, New York: Rinehart.
Polanyi, M. (1967) *The tacit dimension*, New York: Garden City.
Popelard, A. and Vannier, P. (2010) 'America's slow ground zero', *Le Monde Diplomatique*, January. Available at: http://mondediplo.com/2010/01/03detroit (accessed 1 September 2010).
Porter, M. (1998) *On competition*, Harvard, MA: Harvard Business School Press.
Power, D. and Scott, A. (2004) *Cultural industries and the production of culture*, London: Routledge.
Pratt, A. (2006) 'Advertising and creativity, a governance approach: a case study of creative agencies in London', *Environment and Planning A*, 38: 1883–99.
Pratt, A.C. (2000) 'New media, the new economy and new spaces', *Geoforum*, 31: 425–36.
—— (2008) 'Creative cities: the cultural industries and the creative class', *Geografiska Annaler: Series B, Human Geography*, 90: 107–17.
Pruzan, P. (2001) 'Corporate reputation: image and identity', *Corporate Reputation Review*, 4: 50–64.
Rantisi, N. (2002) 'The local innovation system as source of "variety": openness and adaptability in New York City's garment district', *Regional Studies*, 36: 587–602.
Reed, H. (1981) *The preeminence of international financial centers*, New York: Praeger.
Roberts, K. (2004) *Lovemarks: the future beyond brands*, New York: PowerHouse Books.
Sassen, S. (2000) *Cities in a world economy* (2nd edn), London: Pine Forge Press.
—— (2006a) *Cities in a world economy* (3rd edn), London: Sage.
—— (2006b) *Territory, authority, rights: from medieval to global assemblages*, Princeton, NJ: Princeton University Press.
—— (2010) 'Global inter-city networks and commodity chains: any intersections?' *Global Networks*, 10: 150–63.
Saxenian, A. (1994) *Regional advantage: culture and competition in Silicon Valley and route 128*, London: Harvard University Press.
Schoenberger, E. (1988) 'From Fordism to flexible accumulation: technology, competitive strategies, and international location', *Environment and Planning D: Society and Space*, 6: 271–80.
Scott, A. (1996) 'The craft, fashion, and cultural-products industries of Los Angeles: competitive dynamics and policy dilemmas in a multisectoral image-producing complex', *Annals of the Association of American Geographers*, 86: 306–23.
—— (2004) 'Cultural-products industries and urban economic development: prospects for growth and market contestation in global context', *Urban Affairs Review*, 39: 461–90.
Scott, A.J. (1999) 'The cultural economy: geography and the creative field', *Media Culture and Society*, 21: 807–18.
—— (2000) *The cultural economy of cities*, London: Sage.
—— (2008) *Social economy of the metropolis: cognitive-cultural capitalism and the global resurgence of cities*, Oxford: Oxford University Press.

Sirisagul, K. (2000) 'Global advertising practices: a comparative study', *Journal of Global Marketing*, 14: 77–97.

Sklair, L. (2001) *The transnational capitalist class*. Oxford: Blackwell.

Snyder, L.B., Willenborg, B. and Watt, J. (1991) 'Advertising and cross-cultural convergence in Europe, 1953–89', *European Journal of Communication*, 6: 441.

Soja, E., Morales, R. and Wolff, G. (1983) 'Urban restructuring: an analysis of social and spatial change in Los Angeles', *Economic Geography*, 59: 195–230.

Speece, M., Quang, T. and Huong, T.N. (2003) 'Foreign firms and advertising knowledge transfer in Vietnam', *Marketing Intelligence Planning*, 21: 173–82.

Steinmetz, G. (2009) 'Detroit: a tale of two crises', *Environment and Planning D: Society and Space*, 27: 761–70.

Stolarick, K. and Florida, R. (2006) 'Creativity, connections and innovation: a study of linkages in the Montréal region', *Environment and Planning A*, 38: 1799–817.

Storper, M. (1997) *The regional world*. New York: Guildford Press.

Storper, M. and Christopherson, S. (1987) 'Flexible specialization and regional industrial agglomerations: the case of the US motion picture industry', *Annals of the Association of American Geographers*, 77: 104–17.

Storper, M. and Salais, R. (1997) *Worlds of production: the action frameworks of the economy*. Boston, MA: Harvard University Press.

Storper, M. and Venables, A.J. (2004) 'Buzz: face-to-face contact and the urban economy', *Journal of Economic Geography*, 4: 351–70.

Sugrue, T. (2005) *The origins of the urban crisis: race and inequality in postwar Detroit*. Princeton, NJ: Princeton University Press.

Sunley, P., Pinch, S., Reimer, S. and Macmillen, J. (2008) 'Innovation in a creative production system: the case of design', *Journal of Economic Geography*, 8: 675.

Sydow, J. and Staber, U. (2002) 'The institutional embeddedness of project networks: the case of content production in German television', *Regional Studies*, 36: 215–27.

Tallman, S., Jenkins, M., Henry, N. and Pinch, S. (2004) 'Knowledge, clusters and competitive advantage', *Academy of Management Review*, 29: 271–85.

Taylor, P.J. (2001) 'Specification of the world city network', *Geographical Analysis*, 33: 181–94.

—— (2004) *World city network: a global urban analysis*. London: Routledge.

—— (2008) 'Advertising and cities: a relational geography of globalization in the early twenty first century', in Kofman, E. and Young, G. (eds) *Globalization: theory and practice* (3rd edn). London: Pinter.

—— (2009) 'Urban economics in thrall to Christaller: a misguided search for hierarchies in external urban relations', *Environment and Planning A*, 41: 2550–55.

Taylor, P.J. and Aranya, R. (2008) 'A global "urban roller coaster"? Connectivity changes in the world city network, 2000–2004', *Regional Studies*, 42: 1–16.

Taylor, P.J. and Lang, R.E. (2005) 'US cities in the world city network', Washington, DC: Brookings Institution Center on Urban and Metropolitan Policy Survey Series (February).

Taylor, P.J., Catalana, G. and Walker, D.R.F. (2002) 'Measurement of the world city network', *Urban Studies*, 39: 2367–76.

Taylor, P.J., Catalana, G. and Walker, D.R.F. (2004) 'Multiple globalisations: regional, hierarchical and sectoral articulations of global business services through world cities', *Service Industries Journal*, 24: 63–81.

Taylor, P.J., Hoyler, M. and Verbruggen, R. (in press) 'External urban relational process: introducing central flow theory to complement central place theory', *Urban Studies*, 48.

Taylor, P.J., Derudder, B., Saey, P. and Witlox, F. (eds) (2006) *Cities in globalization.* London and New York: Routledge.

Taylor, P.J., Ni, P., Derudder, B., Hoyler, M., Huang, J. and Witlox, F. (eds) (2010) *Urban global analysis: a survey of cities in globalization.* London: Earthscan.

Thrift, N. (1994) 'On the social and cultural determinants of international financial centres: the case of the City of London. Money', in Corbridge, S., Martin, R.L. and Thrift, N.J. (eds) *Money power and space.* Oxford: Blackwell.

Thrift, N. and Olds, K. (1996) 'Refiguring the economic in economic geography', *Progress in Human Geography*, 20: 311–37.

Thrift, N. and Taylor, M. (1989) 'Battleships and cruisers: the new geography of the multinational corporations', in Gregory, D. and Walford, R. (eds) *Horizons in human geography.* London: Macmillan.

Thrift, N.J. (1987) 'The fixers: the urban geography of international commercial capital', in Henderson, J. and Castells, M. (eds) *Global restructuring and territorial development.* London: Sage.

Torstendahl, T. and Burrage, M. (1990) *The formation of professions: knowledge, state and strategy.* New York: Sage.

UNCTAD. (2004) *World investment report 2004: the shift towards services.* Geneva: UNCTAD.

—— (2009) *World investment report 2009: transnational corporations, agricultural production and development.* New York and Geneva: UNCTAD.

Urry, J. (2004) 'Connections', *Environment and Planning D: Society and Space*, 22: 27–37.

US Census Bureau (2007) *County Business Patterns 2007.* Available at: http:www.census.gov/econ/cbp/index.html (accessed 8 April 2010).

Vinodrai, T. (2006) 'Reproducing Toronto's design ecology: career paths, intermediaries, and local labor markets', *Economic Geography*, 82: 237–63.

Von Hippel, E. (1978) 'Successful industrial products from customer ideas', *Journal of Marketing*, 42: 39–49.

—— (2005) *Democratizing innovation*, Cambridge. MA: MIT Press.

Von Nordenflycht, A. (2010) 'What is a professional service firm? Toward a theory and taxonomy of knowledge-intensive firms', *Academy of Management Review*, 35: 155–74.

Wall Street Journal. (2010) 'Chrysler picks four new ad agencies', *Wall Street Journal*, 6 January.

Weller, S. (2007) 'Fashion as viscous knowledge: fashion's role in shaping trans-national garment production', *Journal of Economic Geography*, 7: 39–66.

—— (2008) 'Beyond "global production networks": Australian fashion week's trans-sectoral synergies', *Growth and Change*, 39: 104–22.

Wenger, E. (1998) *Communities of practice: learning meaning and identity.* Cambridge: Cambridge University Press.

West, D. (1987) 'From T-square to T-plan: the London office of the J. Walter Thompson Advertising Agency 1919–70', *Business History*, 29: 199–217.

Whittington, R., Pettigrew, A., Peck, S., Fenton, E. and Conyon, M. (1999) 'Change and complementarities in the new competitive landscape: a European panel study, 1992–1996', *Organization Science*, 10: 583–600.

Wilson, C. (1992) 'Restructuring and the growth of concentrated poverty in Detroit', *Urban Affairs Review*, 28: 187–205.

Wilson, W. (1999) 'When work disappears: new implications for race and urban poverty in the global economy', *Ethnic and Racial Studies*, 22: 479–99.

Wood, P. (2006) 'Urban development and knowledge-intensive business services: too many unanswered questions?' *Growth and Change*, 37: 335–61.

WPP (2008) 'Fast facts'. Available at: www.wpp.com/NR/rdonlyres/28304C33-0E4C-4223–85B8–9889F532996F/0/WPP_Fast_Facts_Jun08.pdf (accessed 18 May 2010).

—— (2010) 'WPP 2009 Preliminary Results'. Available at: www.wpp.com/wpp/press/press/default.htm?guid=%7Be72cf23b-39ee-4fe9-94cd-a8a4fe916670%7D (accessed 11 May 2010).

Yeung, H.W.-C. (2005) 'Rethinking relational economic geography', *Transactions of the Institute of British Geographers NS*, 30: 37–51.

Index

Page numbers in *italics* denote tables, those in **bold** denote figures.

Printed in the USA/Agawam, MA
December 8, 2011

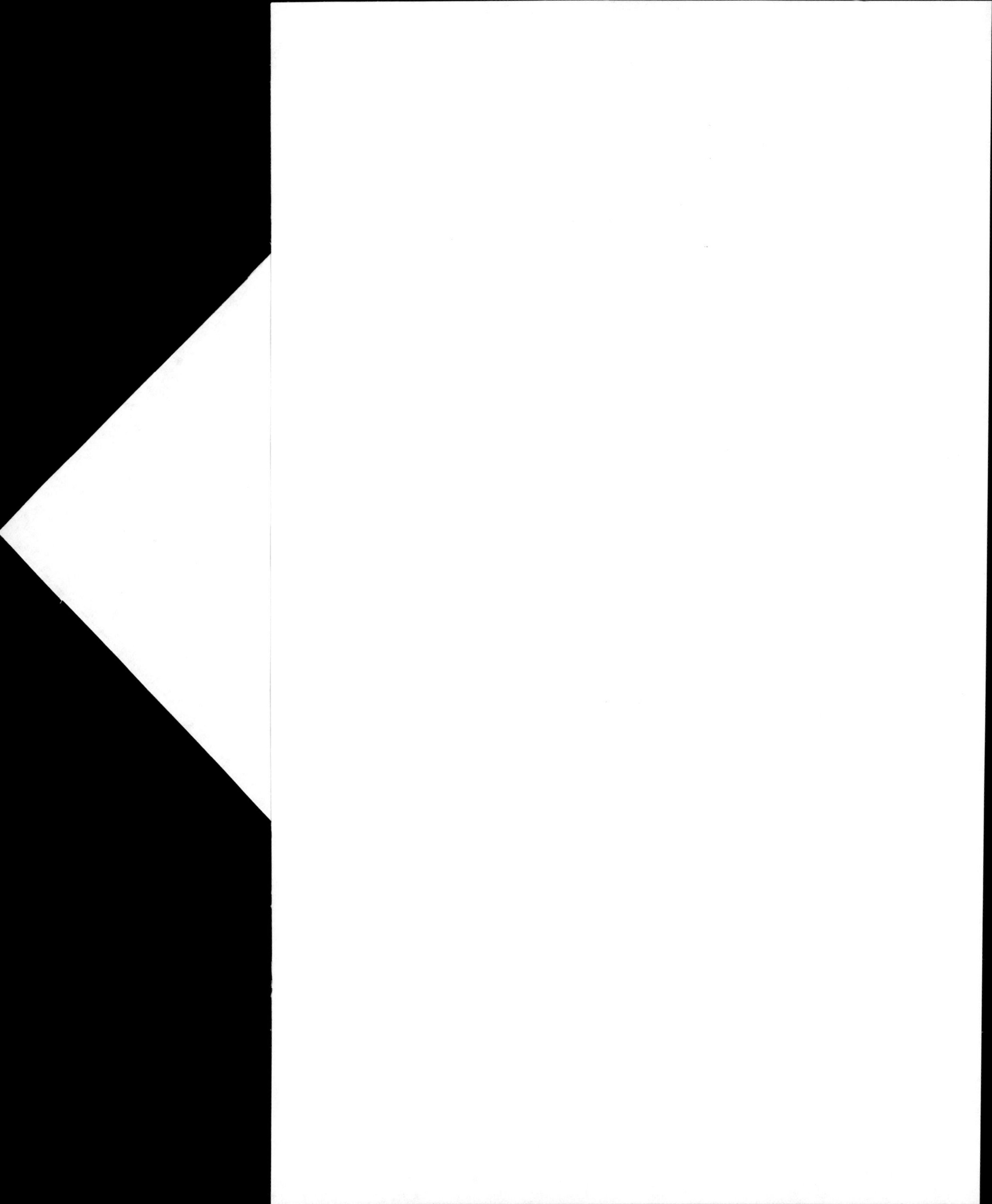